ALL IN A DAY

With grateful thanks to former colleagues,
and Doreen, Dai and Owen

Ill fares the land, to hastening ills a prey,
Where wealth accumulates, and men decay.
Princes and lords may flourish, or may fade,
A breath can make them, as a breath has made.
But a bold peasantry, their country's pride,
When once destroyed, can never be supplied.

– Oliver Goldsmith, The Deserted Village.

ALL IN A DAY'S WORK

Reflections on the FUW & Farming

GWILYM THOMAS

Published by Gwilym Thomas, Aberystwyth, Ceredigion.

ISBN 978-0-9556244-0-7

Printed by Cambrian Printers, Aberystwyth.

Contents

Into the unknown

I had no intention of having anything to do with farming – but to my astonishment I eventually spent 38 years working with and on behalf of the farming community! The Farmers' Union of Wales had been born in a storm of controversy in 1955 – an event which I treated as just another headline and storm in a tea-cup. At the time I was a reporter on a local newspaper and I had no reason to think that my future would be linked to these 'rebels' who toured the country addressing stormy meetings.

I was born and brought up on a small dairy farm on the outskirts of Aberystwyth. I worked on the farm during my teenage years and after two years National Service in the RAF. My father was a tenant of the Nanteos estate and gradually we became aware of the encroaching tide of new houses which slowly surrounded the farm. I had never regarded farming as a career – my interests lay in journalism and when I saw a vacancy on the Welsh Gazette – sadly a casualty of economic pressures some years later – I jumped at the chance. I was taken on as a reporter with special responsibilities for sport. A few years later I joined the Cambrian News as chief reporter and Sports Editor. I enjoyed the work but I was aware that big changes were in the offing in the newspaper and media industry.

Most of my colleagues were being lured away to what were regarded as greener pastures. Some, like Tim Jones (The Times), Harry Pugh (Daily Express) and Robert Ward (Sunday Mirror) went to the national press; most like Tudor Phillips, Herbert Williams and Hywel Jones left for the BBC in Cardiff where the TV companies were expanding their news coverage of Wales.

As I had no desire to leave my home town, one alternative was to provide a news service for the daily newspapers as a freelance journalist. In the two years up to 1964 I built up contacts with journalists and News Editors, particularly in Manchester which, at that time, was the centre for northern editions of the daily newspapers.

The Editor of the Cambrian News, Douglas Wright, was the Mid Wales correspondent for the BBC and when the independent television company 'Television Wales and the West' (TWW) offered me the opportunity of covering the area as a freelance journalist I accepted immediately.

Although TWW expressed their satisfaction with the service, I felt that my news reports lacked one important ingredient in comparison with the BBC – film to illustrate the news items. I took the plunge and, at considerable expense, bought a 16m.m. Eumig cine camera and, for a test period, filmed everything that moved in order to improve my proficiency as a film cameraman.

After a few weeks I decided that I was ready to provide the enhanced service and suggested to the News Editor that I should send him some trial film. It soon transpired that it was not as simple as that and I was told that I would have to become a member of ACTT, the technician's Trade Union which, at that time, wielded considerable power and influence.

Although I was already a member of the National Union of Journalists and, in fact, chairman of the Aberystwyth and Cambrian Coast branch, I duly applied for membership and assumed that acceptance would be a formality. However, I received a reply within a few days informing me that a local freelance cameraman had already been accepted by ACTT only weeks before my application and would be covering the area. It was ACTT policy not to accept membership from another cameraman in the same area. Their decision ended my brief sortie into film – and left me with a superfluous and expensive camera on my hands!

Although I continued to supply news items to TWW and other newspapers there was clearly a limit to this freelance work while I was employed by the Cambrian News and unable to develop the freelance work to its full potential. I realised that sooner or later I would have to choose between the uncertainty of a freelance career and the security of regular, full-time employment on a weekly newspaper.

However there was also a limit to the importance of local events to the national media. Disappointing returns during a holiday period devoted to freelance work only served to strengthen my reservations and fears about a freelance career.

Fortuitously another alternative emerged from an unexpected encounter with the then General Secretary of the Farmer's Union of Wales (FUW), Emrys Bennett Owen. In the autumn of 1964 I embarked on two articles which were to be kept in reserve for the Christmas period. One of these required some background information relating to agricultural statistics and as the FUW headquarters was on my doorstep I sought assistance from the Union. Emrys Owen supplied the information and during my meeting with him it soon became apparent that he was not only grateful for the opportunity to give the Union some publicity but also that the Union was not happy with the publicity which it was receiving from the Press and, in particular, newspapers in Wales.

The Union's relationship with the Welsh media was summed up in a cartoon by J.C. Walker which had appeared in the Western Mail as early as 13th December 1955, a couple of weeks after the union's establishment. It showed a cow – denoted as the NFU – occupying a warm and comfortable cowshed and its calf – 'the new born FUW' – out in the snow and frost of a West Wales winter. The cartoon was headed 'He'll soon come back to mother'. As far as FUW leaders and many Union members were concerned, the cartoon summed up the unsympathetic attitude of the media in Wales. It infuriated FUW leaders at the time and rankled for years – the only consolation was that it made them more determined than ever to survive!

Emrys Owen referred to the fact that the Union's Council had agreed as early as 1957 that the Union should appoint a Public Relations Officer but no action had been taken. A year later a well known agricultural journalist, J.C. Griffith Jones had offered his services but the two parties had failed to agree terms.

As a result, all the Union's Press releases and publicity material was produced by Emrys Owen who freely admitted that despite its importance he was fully engaged in administrative work. In fact, I learned later that the Union was, during that period, experiencing serious financial difficulties and Emrys Owen had the unenviable task of coping with this crisis.

All the Union's Press and publicity material was, at that time, invariably issued after protracted consideration and approval by a committee. The Union's reaction to any development of importance to the farming industry could, as a result, take days and even weeks to emerge. The subsequent delay blunted the impact of the Union's views and I recall suggesting that the problem could be overcome in a variety of ways, not least by cutting out the committee's role and confining any consultative process, if any, to committee chairmen or, better still, the headquarters staff with specific responsibility for those areas. In retrospect this view was rather naive as I failed to appreciate the trenchant views of prominent committee members and their determination that the ultra democratic process which the Union practised and was proud of, should be respected at all costs. I eventually overcame these problems by establishing an understanding and trust with the President of the Union and the system worked effectively during my 38 years with the Union.

I thought nothing of this meeting and conversation at the time but it may well have sown a seed because within weeks the Union decided to advertise the post of Public Relations Officer.

At that time the Cambrian News had a regular farming correspondent, 'Tyddynwr' who had covered the establishment of the Union in December 1955 and its controversial progress. 'Tyddynwr' was a freelance who contributed a regular column on farming. I never knew him and I, and my colleagues in the News Room at Aberystwyth, were more than willing to leave the coverage of the farming industry in his (or her) capable hands. I attended a few stormy meetings of the NFU County Executive at Aberaeron and the official opening of the FUW's headquarters when it moved from its Chalybeate Street site in Aberystwyth to Queen's Square. But these were rare occasions because of my work on major, general news stories and, in particular, sport. Despite my farming background I had a very limited knowledge of the farming industry which discouraged my interest in the subject.

Nevertheless, the FUW vacancy represented a viable, alternative solution to the conflicting attractions of freelance journalism and my regular job. I decided to apply for the post without any real hope of success as the advertisement emphasised the importance of the Welsh language and, at that time my knowledge of Welsh was confined to simple phrases. To my surprise I was offered the post in early December. After a great deal of thought I decided to accept the offer believing that I could always revert to freelance work if I did not measure up to the job or did not like the work.

I did some preparatory work for the Union before Christmas and received an invitation to attend the head office Christmas dinner at a local hotel to meet the staff. I walked into the hotel just before the appointed time and was directed to a member of staff standing by the bar. When I asked where the FUW dinner was he pointed to a back room. As I was walking away he asked if I was the new member of staff. I said I was. I expected some welcoming seasonal remark but to my astonishment his reaction was far from welcoming.

"You'll be sorry," he said, "it's the worst decision you've ever made."

I ignored the remark – it was too late for second thoughts and in the following years, particularly at Christmas, I often thought of those first (un)welcoming words and that episode when, in a state of nervous anticipation I embarked upon a new and challenging career. My ill-wisher was, I learned later, a departing member of staff and I'm glad to say our paths never crossed again.

The Early Years

The Farmers' Union of Wales was established on the 3rd December 1955 at St. Peter's Church Hall, Carmarthen, at the end of a meeting of the County Executive Committee of the rival National Farmers Union (NFU). Ivor T. Davies of Llanfiangel-ar-Arth – 'Ivor Bach' as he was known to his friends and acquaintances because of his diminutive stature – the chairman of the Committee, told the meeting that he had become totally disillusioned with the NFU and that he could no longer support its policies in relation to Wales. He said that he had decided to resign from his chairman's position.

In the subsequent discussion that followed, several members expressed their support for Mr Davies and supported a proposal to establish an independent Farmers' Union of Wales. The decision to proceed with the formation of a new Union had an explosive reaction throughout the farming industry in Wales and beyond but was not unexpected. Earlier in the year Carmarthen NFU had put forward a resolution at the Union's annual meeting advocating that Wales should be accorded the same independent status as Scotland and have a representative on the Union's Price Review team which, at that time, had the vital task of negotiating agricultural policies and financial returns for the industry with the Government. The resolution was heavily defeated.

Ivor Davies and NFU County Secretary J.B. Evans had discussed the possibility of establishing a Farmers' Union of Wales with dissatisfied colleagues from other counties in Wales, particularly neighbouring Cardiganshire where similar discontent had been expressed by leading members of the NFU like D.J. Davies of Panteryrod, Llew Bebb of Goginan and W.J. Capener of Swyddfynnon.

J.B. Evans had discussed the proposal with prominent members in Carmarthenshire and at the end of November another meeting, this time with representatives from Carmarthenshire, Cardiganshire and Pembrokeshire, was held at the Dolwar Hotel, Carmarthen at which there was unanimous support for the proposal.

In March 1955, Lady Megan Lloyd George criticised the NFU for not tackling agricultural problems in Wales and J.B. Evans, whose championship of the cause of Welsh agriculture and family farmers had earned him the admiration and respect of the farming community throughout West Wales, was so disillusioned that he was considering accepting a post with the Country Landowners Association. 'J.B.', as he was popularly known, and Ivor Davies became the driving force behind the campaign to establish an independent Union for Wales.

The scene was therefore set for the momentous decision on the afternoon of the 3rd December 1955. Ivor Davies asked those who were in favour of establishing an independent Union to remain in the hall at the end of the discussion. A number left the meeting but a dozen remained — apart from Ivor Davies they were: Aneurin Davies, D.J. Evans and Tom Price of Cil-y-Cwm, J.W.O. Davies and Aelwyn Hughes of Llandovery, J.H. Davies, Lanybri, Llewelyn Jones of Rhandirmwyn, D.T. Lewis, Myddfai, D.H. Owen, Llanon, D.C. Phillips, Llanelli and Dewi I. Thomas of Ffynnon-drain, Carmarthen.

J.B. Evans had returned to his office but he knew what was happening at St. Peter's Church Hall. It had been agreed that Ivor Davies should take the chair and on the proposition of Dewi I. Thomas, seconded by J.H Davies, it was unanimously agreed that a Farmers' Union of Wales be established. A Provisional Committee was set up with Ivor Davies as Chairman and D.T. Lewis as Vice Chairman.

J.B. Evans was summoned to the meeting and said that he was prepared to accept the position of General Secretary of the new Union on the same terms as he was receiving from the NFU. The Committee agreed to his terms and he was appointed on an annual salary of £1,400.

Just over a hundred years earlier, economic pressures and dissatisfaction concerning government policies had forced West Wales farmers to take a prominent role in the Rebecca Riots. The establishment of the FUW was a peaceful revolution which nevertheless resulted in bitter recriminations and angry exchanges at stormy meetings which split country communities.

The new FUW members and supporters were castigated by a number of NFU branches. They were likened to the Mau Mau, the African terrorists. Families were divided and there were threats to disinherit converts to the FUW cause. Attempts were made to disrupt FUW meetings, particularly at livestock markets and in one incident in West Wales, an incensed FUW supporter nearly came to blows with a prominent local landowner and supporter of the NFU who was later the county's Lord Lieutenant. In Cardiganshire the NFU County Secretary, Bryn Richards described the members of the new Union as traitors and accused J.B. Evans of 'an act of treachery'.

Ivor Davies and J.B. Evans had lit the blue touch paper and the explosive result reverberated through the rural areas of Wales. Both were men of strong principles and as far as J.B. Evans was concerned he had no hesitation in giving up a well paid job with the NFU for the uncertainty of establishing a new organisation from scratch. At a subsequent meeting in Llandeilo of the FUW's Provisional Committee he was instructed to find suitable premises in Carmarthen which would serve as the Union's headquarters.

Within a week J.B. Evans had set up his office at Gwalia House, in John Street, above a car showroom. 'Tyddynnwr' in the Cambrian News reported... 'the entrance is next door to a garage and resembles nothing so much as the vestibule of an old fashioned French hotel. Narrow stairs lead to a suite of empty rooms and in one of these J.B. has his desk, his files and the newly installed telephone. The atmosphere is a cross between a club – with friends popping in and farmers entering to pay their dues – and the offices of a new revivalist organisation. The mood is one of good humour.'

The office was equipped with second hand office furniture and items donated on credit by a local furniture store. J.B. Evans, Ivor Davies and their colleagues were plunged into a round of public open meetings which went on late into the night. In one week, J.B. Evans and others like D.J. Davies, Geraint Howells and Ivor Davies addressed stormy meetings in packed village halls in West Wales and North Wales.

'Tyddynnwr' reported that during his visit to the Carmarthen office, J.B. Evans was making last-minute arrangements for meetings in Merioneth, Caernarfonshire and Anglesey 'before the end of the week and on Tuesday addressed packed meetings at Llanllwni, Llanybyther and Pencader. The following week he was in the Synod Inn and Cwmrheidol areas.'

During this period the NFU poured scorn on the new Union's chances of survival. The NFU's Welsh Secretary, E. Verley Merchant told J.B. Evans "We will smash you in three months". The media in Wales appeared to support this claim – J.C. Walker's cartoon in the Western Mail on 13th December, just a few days after the first meeting of the FUW's Provisional Committee at Llandeilo entitled 'He'll Soon Come Back to Mother' made the founder members even more determined.

Similarly, the Carmarthen Journal was equally unenthusiastic, coming to the conclusion 'It is difficult to imagine with any degree of conviction how a small entity like the Farmers' Union of Wales can ever be able to accomplish more than what can be done by a much larger and more powerful body that is representative of both England and Wales... The sooner the NFU is able to close its ranks the better for the industry and the nation.'

National farming journals were more supportive. The 'Farmer and Stock-Breeder', then an independent journal, commented that the decision to establish the FUW could hardly have come as a surprise. It said that NFU leaders had been too immersed in high politics to detect the trouble that had been simmering. Now, it added, the pot had boiled over. The 'Farmers Weekly' was also critical of the NFU, it reported that the rebellion against the NFU was no sudden flare up. It had been smouldering for years as the returns from milk, pigs and eggs – the small farmer's main products – had been steadily depressed by each Price Review since 1950.

The Cambrian News was also sympathetic – early in the New Year (1956) it reported that the discussion on Welsh farming had never been so stimulating. Its correspondence columns, it added, had carried letters of extraordinary frankness. When its columnist visited FUW headquarters, J.B. Evans received a phone call from a 'Farmer and Stock-Breeder' reporter asking for an interview. The reporter was invited to attend meetings in the area and when he went to a meeting at Llanybyther he found the hall packed with people, many sitting on the window sills – the speakers had difficulty in getting into the meeting and on to the platform.

In his article in the 'Farmer and Stock-Breeder' the following week he was enthusiastic about the support for the new Union. "The force of the movement should not be underestimated" he wrote. "Officials of Carmarthen NFU say that they have no fear... they should not be too confident. The new movement has made a good start and if it progresses in the same way it will gain ground rapidly."

Four years after its establishment – in May 1959 – support came from an unexpected source. In a glowing commendation the Financial Times stated: "The FUW is a force to be reckoned with... its roots go down into deep soil, invigorated, as it may be, by past frustrations and controversies, but fed principally from the conviction that the Welsh voice can do more for Welsh agriculture solo than in chorus. To attribute the movement to Welsh nationalism is to mistake symptoms for causes. As with Welsh nationalism, so with the FUW – in its moderate form, it is the expression of a native spirit, of a proprietary branch of culture that conceives of unity on a national basis in the only sense that unity has value, as federation from conviction rather than amalgamation for the sake of convenience... The FUW has the ball at its feet. It is only a matter of time before it wins, for itself and its principles, the recognition it has sought from the start. It is here to stay... As a new farmers movement it will be regretted only by those who see power as monolith, and influence in terms of the anonymity of large numbers."

Despite this support and encouragement, and despite the initial flush of success the immediate future was not without its problems. A number of well-meaning individuals rose above the exchange of bitter recriminations and attempted to pour oil on troubled waters. One of the first to do so was Captain J. Hext Lewes in Ceredigion, a leading member of the NFU who pleaded with the NFU hierarchy to reach a compromise with the new Union. His appeal fell on deaf ears. Another was J.J. Davies of Llanina Mansion, New Quay, also in Ceredigion whose efforts to heal the breach in December 1957 precipitated a crisis of confidence in the leaders of the FUW, particularly J.B. Evans and D.J. Davies who had, as a result of D.J. Davies' efforts, met NFU President Sir James Turner and other NFU leaders in London, to discuss the possibility of unity without the knowledge of other FUW leaders. Ivor T. Davies was particularly critical and although he considered that J.B. Evans had fallen into a trap, the unauthorised meeting undermined J.B.'s position. The fall-out from this London meeting was so severe that it inhibited any immediate prospect of unity and engendered a feeling of mistrust for years to come.

FUW members in Brussels in 1980.

In fine voice – FUW visitors to Brussels pass the time on the ferry by bursting into song.

Gwilym Thomas with Barry Alston (Farmers' Guardian), John Phillips (FUW Deputy President), Emyr Jones (Rhandirmwyn), Derek Morgan (FUW), Bob Parry (FUW President), at Clynmawr, Rhandirmwyn.

With Merioneth members and staff at the National Eisteddfod.

French TV crew in Mid Wales when shops boycotted French goods.

The Welsh Agricultural College's soccer team in shirts sponsored by FUW insurers NEM. The picture includes Peter Hutton and Alan George (NEM) and H.R.M. Hughes and Gwilym Thomas (FUW).

The late Norman Fitter, the Union's Executive Officer for many years, being presented with a retirement gift by General Secretary Evan Lewis.

Minister of Agriculture Nick Brown with President Bob Parry and Helen Lloyd after opening the Union's new HQ.

Ceredigion FUW's contribution to the National Eisteddfod at Aberystwyth in 1992.

An effective partnership – Myrddin Evans (President) and his Deputy H.R.M. Hughes, seen here at a Brussels briefing.

Early leaders, the late Glyngwyn Roberts, Myrddin Evans and H.R.M. Hughes.

First General Secretary J.B. Evans, first female Vice President Megan Davies, and Anglesey's Secretary, the late R.J. Williams.

Mid Wales pioneers, the late Lord Geraint, former Deputy President R.O. Hughes, Rhydwyn Pughe, Merioneth (Vice President).

Union's Aims & Objectives

One of the first tasks which faced the founder members of the Union was that of listing its aims and objectives and the reasons for its establishment. The list was not difficult to compile as it included grievances and concerns which had been aired at numerous farmers meetings, including those of the NFU. The task was undertaken by J.B. Evans, Ivor Davies and Deputy President D.T. Lewis. They eventually compiled the following eight points:

1. *Existing organisations had failed to safeguard the interests of the farmers of Wales.*
2. *Small farmers had no representation on the Price Review team.*
3. *Wales was refused representation on the Price Review team.*
4. *Wales was refused two representatives on the Wool Marketing Board.*
5. *Wales was refused two representatives on the Egg Marketing Board. Northern Ireland had four representatives and Scotland two representatives.*
6. *Wales was treated as a region on all producer marketing schemes because it did not have a Farmers Union of its own.*
7. *An application to the NFU for a conference on problems affecting Wales had been refused.*
8. *The increase in the NFU's subscription rates favoured the big farmer; the member farming 85 acres was being asked to pay the same increase as the member farming 2000 acres.*

The Union's rules were not finally formulated until nearly two years later, in April 1957. The ten points outlined the aims and principles of the Union which were summed up in the opening paragraph:

To promote the interests of those engaged in farming and agriculture in Wales by facilitating the cooperation of farmers and others interested, with a view to the protection of agriculture as an industry by furthering legislation which will, in the opinion of the Union, be sound and beneficial, and by opposing legislation which would, in the opinion of the Union be prejudicial to the interests of farmers or agriculture, and by any other lawful ways and means which, in the opinion of the Union, shall be conducive to the attainment of this object.

The rules also emphasised the Union's willingness to cooperate with other organisations with similar aims in the best interests of Welsh agriculture.

In 1964 the Union made minor revisions to its aims and objectives by publishing a pamphlet 'Why the Farmers Union of Wales?' This listed ten points:

To attain for the farmers of Wales, equal status with their fellow farmers in England, Scotland and Northern Ireland through their respective Unions.

To cooperate fully and equally with other farming organisations in the United Kingdom and overseas.

To safeguard the interests of all farmers in Wales by pressing for:

Full and fair representation on all Marketing Boards and statutory bodies affecting agriculture in Wales.

The cost of production on Welsh farms to be independently considered and farm incomes in Wales to be brought up to a realistic level in comparison with those in other industries and other countries in the UK.

The establishment of two-tier price systems of payment for milk and eggs.

The provision of one producer, one vote voting systems for the Milk and Egg Marketing Boards.

The setting up of a producer controlled Meat Marketing Board.

The maximum share of the home market for home producers in all commodities.

The equitable allocation within the industry of payments and grants made by the Exchequer.

In 1965, Union President Glyngwyn Roberts encapsulated the general principles and aims of the Union when he stated at the annual meeting:

"The aim of this Union is cooperation on an equal status with our fellow-farmers in the other home countries. We seek the right of free determination, the right that has been so long enjoyed by farmers in England, Scotland and Ulster but which has been denied to Wales... From its very instigation the Farmers' Union of Wales has sought to create and introduce policies which would benefit the majority of farmers in Wales – to create an independent pattern to the suit that the industry has worn since the 1947 Agriculture Act. We live in a changing world in which change is not always synonymous with progress. Farmers, big and small, could play their part in this country's economic balance sheet... they should be given the opportunity of doing so."

In 1973, FUW President Myrddin Evans was able to pin-point areas where the Union's influence had paid dividends. He referred to the successful opposition to the drowning of the Gwendraeth valley, and the establishment of a Welsh regional office for the Intervention Board which had not been included initially in the government's proposals. FUW staff had succeeded in obtaining enhanced pipeline easement payments in North Wales and improvement in brucellosis compensatory payments – in South Wales pioneering progress had been made in solving sheep straying problems. The Union had campaigned successfully for the establishment of the Welsh Agricultural College at Aberystwyth – a campaign which had mooted the importance of siting a new veterinary college in the town. The town's University had offered a 400 acre site for the project at the time and told the Swann Committee that it should be included in its development plan. The proposal, initiated by the Union's Montgomeryshire branch, received wide support but has still to come to fruition.

Mr Evans highlighted the importance of the Crowther/Kilbrandon Commission report on devolution and added: "We are on the threshold of an exciting and inspiring future in which we hope that the report will set the seal on some form of elected assembly for Scotland and Wales."

Some years later FUW President H.R.M. Hughes also looked back on just some of the Union's achievements – from the Union's early influence on the government's Small Farm Scheme, its successful opposition to the Rural Development Board with its compulsory powers in Mid Wales, its success in mounting opposition to the drowning of valleys in Mid Wales for reservoir purposes and its opposition to a new National Park in Mid Wales.

Mr. Hughes added that when the FUW was established the Price Review procedure was all-important in determining the well-being of Welsh farming. Unfortunately, Welsh farming had no independent representation to safeguard its interests, no independent input into the Review, while at the same time, it did not have a government minister with responsibility for Welsh agriculture. All that changed in the following years and one would be hard put to deny that the Union had no influence on these events or the progress towards devolution. The Union has proved to be a powerful and effective guardian of Welsh agriculture and rural Wales.

The Pioneers

The founder members and pioneers of the new Union were typical of the family farmers of West Wales at the time. They were brought together by a common cause and inspired by strong principles and a burning desire to rectify the injustices which threatened their viability and very future. A small band of leaders emerged who worked their farms by day and then travelled to all parts of Wales to spread the FUW gospel.

J.B. Evans was, in my opinion, the inspiration, and most influential figure in the establishment of the Farmers' Union of Wales. He was one of those rare characters – a man of courage who was prepared to sacrifice his own interests for a principle. I had the good fortune to know and work with J.B. – as he was known – when he returned to the Union as Carmarthen FUW's County Secretary for four years from 1966 to 1970. Some time later I recorded a conversation with him after he had retired and was living near Llanybyther. Much of that interview appeared in the Union's journal 'Y Tir & Welsh Farmer', of which I was the Editor, in a special supplement marking the Union's recognition by the government in 1978.

The seeds of rebellion had undoubtedly been sown as early as 1949 when the Forestry Commission put forward a proposal to take over 20,000 acres in the Rhandirmwyn area. The London headquarters of the NFU approved the scheme but J.B. castigated the proposal and received the backing of his county branch which fought for the 40 or so farms which would have disappeared had the scheme gone through. The London NFU's failure to oppose the scheme resulted in a great deal of bitterness in Carmarthenshire. Carmarthen NFU's opposition was vindicated when the incoming Conservative government scrapped the scheme.

There were numerous other reasons for the increasing disillusion with the NFU, most of it stemming from the fact that the founder members of the FUW considered that unlike Scotland and Ulster,

Wales had little or no influence on policy decisions affecting vitally important areas like the Price Review and the Marketing Boards which, in the post-war era, had a significant influence on returns from commodities like milk and wool, which were of vital importance to Welsh farmers. J.B. and his founder-member colleagues considered that the Welsh Committee of the NFU was largely powerless to safeguard the interests of Welsh farmers. J.B. was a saddler's son and had no practical experience of farming but he saw at first hand the privations of the small, family farmers in West Wales.

At the outbreak of the second world war he volunteered as an ambulance driver in the Army and went to France with the British Expeditionary Force. In 1940 he was commissioned in the RASC and posted to the Middle East and served in Libya and the Western Desert. He was taken prisoner by the Germans during the courageous defence of Tobruk and was a prisoner of war in Italy, Bavaria and Germany until the end of the war. After the war he was appointed Clerk and Chief Finance Officer to Newcastle Emlyn Rural District Council, a post he held until December 1947 when he was appointed County Secretary of the Carmarthenshire branch of the NFU. In the following eight years his influence on the local farming community can only be described as remarkable, and he increased membership of the branch from 1,500 to over 5,000.

In 1952 be was selected as the Conservative party's candidate for the Parliamentary constituency of Brecon and Radnor. At the 1953 Conservative party conference he sponsored a resolution criticising the farming policy of the party. He denounced the party's failure to produce policies to safeguard the family farm. When the resolution was defeated he promptly resigned his candidature. In 1959 he was persuaded to stand for the Carmarthenshire constituency but was unsuccessful. In accordance with the Union's strict code of political neutrality he had resigned his position with the FUW and subsequently held administrative positions with the Conservative party in Merioneth, Montgomery and Neath until he finally retired at the age of 60. Later he responded to a plea from the FUW and returned as County Secretary of the Carmarthenshire branch.

At a presentation ceremony to mark his final retirement at a 15th anniversary celebration of the Union at Carmarthen, the then President of the Union, T. Myrddin Evans, paid tribute to his services to Welsh agriculture in general and Carmarthenshire farmers in particular. Of his contribution to the Union, Mr. Evans said:

> "No one pursued the FUW crusade with more determination, enthusiasm and zeal than J.B. He abandoned a secure position in an established organisation for a principle. One can only guess at the enormity of the burden he carried in those early days."

Concern about the future of the family farm was uppermost in J.B.'s mind when I recorded the interview with him at his home near Llanybyther in 1970. His thoughts and strong views were remarkably perceptive when one looks back at the end of the century. At that time – in the early 1970s – the farming industry was going through a very difficult period – indeed it was not long before frustrated farmers resorted to militant action in some areas of Wales.

J.B. laid the blame on the policies of successive governments which had paid lip service to the needs of the industry. When it came to practical support, every Price Review during the 1960s had, without exception, resulted in under-recoupment. The industry had been under capitalised, resulting in contraction and not expansion.

He prophesised that unless governments recognised the danger, the future of the family farm would be put at risk and much of the character of rural Wales would be destroyed – social and cultural values would be undermined. He criticised government schemes to amalgamate farms and warned that increasing environmental and recreational pressures would inhibit farming in some areas. He was also concerned about the problems facing young farmers and urged central government and local authorities to provide special incentives to encourage entry to the industry.

J.B. was opposed to entry to the Common Market, fearing that it would be difficult to achieve a level playing field. He was afraid that the needs of Wales would be drowned in the sea of conflicting national interests on the Continent – he said that he would only approve of entry if Wales had avenues of independent representation.

He added: "This view is not a nationalistic one but is based on 30 years experience of agricultural administration. For government purposes, Welsh agriculture has always been taken with that of England and that is where the trouble starts because Welsh priorities may not always be shared by those of England, Scotland and Ulster. Agriculture in England is governed by the large-scale farmer and for the last ten years agricultural policies have been formulated to assist large farms."

Turning to the milk industry – the all important sector for West Wales farmers – he said he had fears about the future of the Milk Marketing Board. His fears became reality some years later. Referring to claims of over-production he said that no one could convince him of over-production when the UK was importing thousands of tons of butter and other milk products from overseas. Many of the contracts were for 10 to 15 years whereas home producers had to contend with the uncertainty of a twelve-month policy review.

He prophesied that in a increasingly competitive market, the marketing and promotion of commodities would have to be improved.

J.B. was only reticent on one subject – that of his untimely and controversial departure from, the Union in 1959. It was clear that the memories of that period were still painful and distressing; he felt that his intentions had been misunderstood and that he had been unfairly criticised by some leading members of the Union. The episode is well documented, and involved an unauthorised meeting in London between J.B. and D.J. Davies and the President of the NFU, Sir James Turner.

Two years after the establishment of the Union – in 1957 – some of the original animosity had declined and tempers had cooled. Some well-meaning and respected individuals considered that the breach in farming representation was not in the best interests of the industry and made honest attempts to bring the two sides together.

One such individual was Col. J.J. Davies of Llanina Mansion, New Quay in Ceredigion who had approached D.J. Davies and later Sir

James Turner urging them to end hostilities and heal the breach. Sir James indicated that he would have to put the proposal to the NFU's Welsh Committee which met on December 16 in London. A deputation from the FUW happened to be in London on the same day to meet representatives of the Union's insurers, National Employers Mutual.

After the meeting the President, Ivor Davies and Deputy President, D.T. Lewis decided to go to the House of Commons to listen to a debate. J.B. left to take his brief case back to his hotel, fully intending to rejoin his colleagues later. Back in the hotel however, he met D.J. Davies who told him that Sir James Turner wanted to see them urgently. They were given to understand that it would be a private meeting but when they got there they were surprised and angered to discover four senior members of the NFU Welsh Committee with Sir James. D.J. Davies read out a prepared statement on unity but it soon became clear that Sir James and his Welsh colleagues were not prepared to make any concessions.

When J.B. and D.J. Davies informed Ivor Davies and D.T. Lewis the next morning what had taken place both were furious and J.B. was instructed in no uncertain terms to call a special meeting of the Union's Council. At that meeting and subsequent stormy meetings early in 1958, J.B. and D.J. Davies tendered their resignations and although they were not accepted initially, J.B. was finally replaced by W. Phillip Davies, former Town Clerk of Aberystwyth, but continued as the Union's Agricultural Adviser. A year later he resigned to fight the Carmarthenshire constituency for the Conservative party. Although he continued to have some contact with the Union after his failure to win the seat, his standing was not fully restored until he responded to an appeal from the Union to return as Carmarthenshire's County Secretary some seven years later.

D.J. Davies later claimed that he only knew that Welsh NFU representatives would be present at the meeting shortly before-hand. J.B.'s unauthorised presence at the meeting was so unusual and so out of character that I pressed him on this point. He agreed that they had not known until the last minute that Sir James would not be alone but he finally admitted that he was driven by a much more important reason – a strong indication that the NFU President was prepared to accept Wales' independence within a federated UK structure of representation.

J.B. hinted that his source for this information was Colonel J.J. Davies who was a leading member of the NFU and had access to Sir James Turner. J.B. said that he went to the meeting confident that the Union's founder members had achieved their objective and that they would be able to return from London in triumph.

"It was too good an opportunity, it was a risk, but my informant was so convinced that we would be granted the same status as Scotland that I believed the risk was minimal. Sir James was said to be concerned about the damage that the rift was having on the industry."

However J.B. underestimated the opposition and hostility from Welsh Committee members of the NFU and Sir James had second thoughts about acceding to FUW demands. J.B. emphasised that having played such a leading role in the establishment of the FUW and sacrificed his position with the NFU he would never have undermined the FUW's position. If the FUW was to achieve its objective someone would, some day, have to sit around the table with the NFU.

J.B. also said that he had attempted to contact Ivor Davies and D.T. Lewis, the Deputy President at the House of Commons but without success.

J.B. claimed that at the time the NFU headquarters had become seriously alarmed about the exodus of members to the FUW – so alarmed that a number of tentative approaches about unity had been attempted. But the hostility of some members of the NFU's Welsh Committee was so strong that they came to nothing and the storm which erupted inside the FUW after the London meeting was such that there was little hope of success. Col. J.J. Davies became so frustrated by the NFU's prevarication that he resigned his membership and joined the FUW. But he continued to advocate unity and a few months later he set out another proposal to integrate the NFU and FUW. Again it came to nothing.

J.B. left the Union under a cloud not long after the London meeting – incredibly some of his critics overlooked the enormity of the inital sacrifice he made when he left the NFU to establish the FUW.

When he established the Union's administrative structure he was faced with the mammoth task of finding and equipping an office and building an organisation which had no monetary means of support except for the aspirations of a small group of farmers, nine of whom contributed a £50 loan to overcome initial costs. But the early days were fraught with difficulties and. one doubts whether anyone with less determination and courage than J.B. would have succeeded. His efforts were finally recognised when he was made a Life Member of the FUW Council, an honour which finally convinced him that he had been forgiven and his contribution fully recognised.

Ivor T. Davies, the first President of the FUW farmed at Brynmafon, Llanfiangel-ar-Arth and had been prominent in local government and farming circles before being elected as Chairman of the Carmarthenshire branch of the NFU. Known as Ivor Bach because of his short stature, he more than made up for his lack of inches by a fiery and steely determination to defend the interests of his fellow farmers. A plain speaking individual he would brook no nonsense from his critics and detractors; he forged a close association with his County Secretary, J.B. Evans. Both had the same strong views on fruitless journeys to NFU meetings in London and it was typical of both that they decided that something tangible should be done about it.

Ivor Davies was undoubtedly shaken by J.B. Evans' infamous meeting with Sir James Turner in 1957 and it was typical of him that he demanded an immediate, full-scale and open inquiry into the circumstances of the meeting irrespective of his close association with J.B. I have no doubt also, that Ivor Davies' decision to step down from the Presidency a year later – after just two and a half years in office, was strongly influenced by that unauthorised meeting. Although he was subsequently co-opted on to the Council and all headquarters committees, he distanced himself from the Union and became less actively involved in its work.

His strong views on any attempt to come to terms with the NFU was illustrated by his attitude to a personal appeal for unity made by a prominent Glamorgan member of the NFU at the 1960 annual meeting of the Carmarthenshire branch of the FUW at Llandeilo. Despite his impartial and conciliatory appeal his presence was condemned by the former President who said it was a fatal blow for an NFU member to address the branch. He added pointedly: "I cannot understand the branch for requesting a member of that nature here. A member like that is no good to us at all. We don't want him. We don't want members of the NFU to address us . I hope you won't do this again."

However despite his apparent coolness towards the Union in those immediate years after he stepped down from the Presidency he was delighted to join his founder member colleagues at Aberystwyth for the tenth anniversary dinner and braved inclement weather to get there. He was also proud to be elected a life member of the Union's Council and joined other former Presidents in a happy reunion in 1994 and 1999 in Aberystwyth when the Union opened its new headquarters.

D.J. Davies of Pantyryrod, near Aberaeron was an influential and controversial founder member of the Union and played a leading role in its establishment. He was vice-chairman of the NFU's Welsh Committee and had been nominated for the vice Presidency of the Union. Many tipped him as a future President of the NFU. One leading member of the NFU, Mr N.S.K. Pugh, lamenting his loss to the Union during a meeting at Aberaeron, said that his departure ended the hope that they would have a leading official from Wales.

D.J. Davies was a natural leader in all aspects of his life – a sometimes impetuous character, he was never far from controversy but earned the respect and admiration of fellow farmers by his energy and constructive views on the future of the industry.

'D.J.' as he was known, emerged from a modest background. The son of a Ceredigion small farmer, the family moved to Llangeler and D.J. attended the secondary school at Llandyssul. His father died at an early age and he had to leave school to help his mother on their small farm. Soon he was running two farms and was a prominent member of the YFC. When his club, Bwlchygroes, won the National Efficiency shield. D.J. produced a report on the competition which was published for national and international circulation. He was, for six years during the second World War, employed as an animal husbandry officer by the County Council before taking over the then 270 acre Pantyryrod farm near Llwyncelyn. He developed the farm into one of the most modern and productive farms in the country despite the fact that he was very much involved in a variety of organisations and committees connected with farming. A socialist all his life he was chairman of the Ceredigion Association and contested the Parliamentary seat, paving the way, some believe, for the party's eventual victory by Elystan Morgan.

D.J. was vice chairman of the Welsh Committee of the NFU and a member of the Union's Council when he became involved in the establishment of the FUW. At an NFU meeting in December 1955 he was asked to swear allegiance to the Union. He replied that he could only swear allegiance to the Ceredigion farmers who had elected him as their representative. He added that he supported the FUW and he and Llew Bebb, the other Ceredigion representative, then walked out of the meeting. In a letter to the President of the NFU, Sir James Turner, he outlined his reasons for leaving the NFU and appealed to Sir James to consider the views of those who supported an independent Welsh Union. He made another plea for common sense to prevail and added "I wonder whether it is too late for wisdom to replace unbridled fury."

In 1956 the FUW produced its first major policy statement, 'Draft Long Term Policy for the Small Farmer' which was drafted by D.J. with the assistance of J.B. Evans and Assistant Secretary John Morris, a future Secretary of State, who later attained the highest honours in legal and political circles.

The policy statement warned that unless government policies changed, small, family farms would face extinction. It included a five year development plan for the industry, advocating low interest loans and cheap credit facilities for small farmers. It gave prominence to the problems of milk producers and suggested a two-tier system of payment favouring the small producer. Price Reviews should be held every five years and not annually in order to introduce a measure of long term stability. Two of the main proposals relating to the annual review and modernisation grants were subsequently included in a government White Paper – 'The Long Term Policy for Agriculture.'

A number of other recommendations emerged from a subsequent Policy Conference, including the call for the establishment of a Welsh Agricultural College – which came to fruition some years later – and advocating the establishment of a Welsh Veterinary College – still currently advocated by a number of rural organisations.

The Union's emphasis on the plight of small farms was finally recognised by the government when it issued a White Paper on assistance for small farmers and the Small Farmers Bill which set out a system of differential payments.

In the meantime D.J. had resigned from the FUW over the NFU London meeting but he was back in 1960 when he produced another policy document on the advantages of group farming.

D.J. was an effective speaker and I once saw him reduce a meeting to uproar with his caustic remarks. When he stood on his feet the room would fall silent and D.J. would gradually build his case until he had his listeners in the palm of his hand. He had an unruly lock of hair which refused to lie down and be disciplined – a reflection of his own personality. He reminded me of Aneurin Bevan as he made his points by jabbing his finger in the air. He infuriated his critics, in particular another prominent, early member of the Union, Captain H.R.H. Vaughan of Rhandirmwyn, a former Naval officer who had spent his life living within the confines of Naval discipline. He was the chairman of the Union's Policy Committee and resigned over D.J.'s role in the London NFU meeting. He cursed D.J.'s influence on Council meetings – "He turns up twice a year and each time makes a pestilential nuisance of himself and frustrates business until he gets what he wants."

D.J. was an enigma – promoting the interests of small farms as he developed his own farm into one of the largest and most efficient dairy farms in Wales. He accepted an invitation to serve as a member of the Mid Wales Rural Development Board which the FUW vehemently opposed on the grounds that it had compulsory powers to amalgamate small farms.

One of D.J.'s last victories was at the 1967 annual meeting of the Union when there was serious concern and uncertainly about the effect of Common Market entry on Welsh agriculture. The leader of the Liberal party, Jeremy Thorpe MP was the guest speaker and emphasised the advantages of UK entry. However the meeting succumbed to D.J.'s ringing denunciation of membership when he warned that Welsh rural life would be decimated and that empty farms would stand like tombstones on the hillsides of Wales. Much to Jeremy Thorpe's chagrin the meeting overwhelmingly backed D.J.'s opposition to entry.

D.J.'s mischievous character was never more clearly illustrated than the time he burst into my room at FUW head office one day and described an astonishing encounter with a senior official of the NFU. The NFU official had, the previous week, met a mutual acquaintance of D.J.'s at a farming event in the south of England and had mentioned that he was travelling to an Aberystwyth meeting the following week. He was told that D.J.'s farm lay just off his route near Aberaeron and that D.J. would be interested in meeting him despite his links with the FUW.

The official had duly called at the farm and was soon immersed in a lively discussion on farming representation, Union finances and county administrative structures. To D.J.'s amazement his visitor illustrated a point by producing a large ledger with facts and figures reflecting the NFU's membership and administrative structure in Wales.

D.J. said that his eyes nearly popped out of his head as he took in the detailed information laid out in front of him. He chuckled as he recalled that as soon as his visitor left he had rushed for pencil and paper to record as many details as he could remember. The incident was all the more remarkable in that the encounter took place when there was considerable controversy and speculation in Wales about the respective strengths of the Unions in the media. It is a tribute to D.J. that he never took advantage of the official's actions or mentioned the incident publicly. Nevertheless we found some of the information he gleaned extremely interesting!

As early as February 1956 the Cambrian News farming correspondent, Tyddynwr, recognised D.J.'s qualities and described him as one of the best spokesmen rural Wales had produced. Another commentator summed up DJ.'s ability and mercurial character by quoting the Welsh poet, Dyfed:

'Dyn ar dan, er daioni, y mae grym a gwres yn gwreichioni'

which, roughly translates to:

'A man of integrity and passion for good who illuminates and inspires'

Glyngwyn Roberts was the first of a crop of Union leaders from Anglesey. From its early days he was one of the most influential and highly respected members of the Union and was venerated for a considerable period, first as an outstanding and charismatic leader and later as an influential voice at local and national level. Anglesey was one of the first counties to support the new Union in March 1956 and Glyngwyn Roberts was elected vice-chairman of the first county committee. His ability was soon recognised throughout Wales and he became a Vice President in 1956, Deputy President two years later and President in 1961 – a meteoric rise through the ranks!

A native of neighbouring Caernarfonshire, he had moved to Anglesey where he soon gained a reputation for his progressive farming methods. A former member of the NFU, he took an active interest in local affairs, first as a prominent member of the YFC and later as a member of the local district council. He was an accomplished speaker in Welsh and English; he recognised the importance of publicity and willingly cooperated in promoting the image of the Union. He reflected the virtues and values of a rural and non-conformist Wales and he gave no quarter in his commitment to the Union's policies and objectives.

Glyngwyn Roberts was incensed when he heard that J.B. Evans and D.J. Davies had attended the London NFU meeting and he played a leading role in condemning their actions and any suggestion that the new Union should compromise its ideals. He came to represent the views of North Wales branches during the difficult early days and grew in stature as the Union progressed.

Glyngwyn Roberts was the first President I worked with and one problem I encountered was his assumption that my knowledge of the Welsh language was equal to his. Nothing could be further from the truth and I often thought that the services of an interpreter would have been useful! From the outset I introduced a regular flow of Press statements on a variety of issues, in particular his addresses to county branches and his – and the Union's – reaction to government policies and issues relating to the industry. This necessitated a prior discussion – usually on the telephone – on what he intended to say. His response was usually a torrent of Welsh in Anglesey patois which was beyond my understanding! My response was to guess what he was saying and send him a copy of the Press statement – sometimes at the last minute. He never objected to the contents and we perfected the system into a fine art.

Unfortunately Glyngwyn Roberts' Presidency coincided with an alarming deterioration in the Union's finances. It had responded positively to the demand for new offices and staff but then found that the first flush of success had subsided in membership terms. He and General Secretary Emrys Owen faced a number of financial crises which developed to threaten the very future of the Union. It was not surprising that Emrys Owen left the Union to establish an estate agency in Merioneth in the Spring of 1966 and that Glyngwyn Roberts – who had little interest in the minutiae of administrative affairs – relinquished the Presidency to contest the North Wales Milk Marketing Board regional election.

Marketing Board elections were regarded as a test of strength of the two Unions as, in nearly all cases, the seat was either held or contested by NFU nominated candidates. Glyngwyn Roberts' opponent was the sitting candidate and well known NFU member, John Evans of Machynlleth. At that time both Unions recruited canvassers and collectors throughout the region and the election campaign was run like a military operation with the outcome dominating the agricultural scene.

The FUW had always complained that its candidates suffered under the electoral system of marketing boards at that time as it was weighted in favour of large-scale producers. In Milk Marketing Board elections, producers were entitled to one individual vote and one vote for every ten cows in their herd.* With the possible exception of Anglesey, the FUW represented small-scale producers and as a consequence the NFU candidate was favoured to retain the North Wales seat.

The result was a victory for the NFU candidate, John Evans, who polled 7,401 votes to Glyngwyn Roberts' 6,601 votes, a majority of 800 votes. The result, regarded as a close call by independent observers, and by the FUW as another example of the inequality of the voting system, was challenged by Glyngwyn Roberts who alleged that some collectors had not submitted voting papers marked in his favour to the Board and had not been included in the count. He obtained sworn affidavits from some producers who had voted for him and he asked the Board to trace their votes to ascertain whether they had been recorded.

Farming News commented at the time: "We are quite sure no canvasser on either side would be so criminally stupid as to deliberately destroy or withhold papers."

Glyngwyn Roberts and the Union did not share the view and the doubts were strengthened by the Electoral Reform Society which stated: "We have found much that is unsatisfactory and undemocratic about the Board." It described the 'cow vote' as 'excessive'.

Glyngwyn Roberts was determined to pursue the issue and he asked his MP, the then Secretary of State, Cledwyn Hughes, to press for an inquiry. Cledwyn Hughes, later Lord Cledwyn, decided that there was nothing that he could do and as a result, the already cool relationship between the two, intensified on Glyngwyn Roberts' part.

The Milk Board election, and its aftermath, played a part, I believe, in the Union's decision to publicly snub Cledwyn Hughes

---

* The MMB's voting system was not reformed until 1986 when a one-man-one-vote system was introduced. In the first subsequent election for the South Wales seat FUW nominee Roger Evans defeated NFU candidate Harold James.

when he was Minister of Agriculture in 1970. When the Anglesey MP was elevated to his Ministerial role at the Ministry of Agriculture there were high hopes in the FUW that he would officially recognise the Union – that is, formal acceptance by the government of the FUW's position and role as a recognised representative of producer opinion and, accordingly, its statutory right to speak to the Minister of Agriculture and his officials, on behalf of farmers in Wales. FUW President Myrddin Evans had been the first leader of the Union to meet the Minister of Agriculture, then Fred Peart, at Aberaeron but Cledwyn Hughes was the first Minister to officially visit the FUW stand at the Royal Welsh Show in 1968. Significantly the main topic was the government intention to share agricultural responsibility between the Minister and Secretary of State.

FUW leaders were confident that recognition – which had been the Union's goal since its establishment – was around the corner, particularly when the Minister accepted an invitation to address the Union's annual general meeting in 1970.

The outcome of the 1970 Price Review was however a disastrous let-down for the industry which confidently anticipated that the government would tackle the recession and restore some semblance of confidence. Instead, the Price Review settlement was regarded as the worst for years and most of the blame was laid at the Minister's door. Angry delegates at the five-hour Council meeting called to discuss the Review were in a militant mood and were told by Myrddin Evans that the industry had, in monetary terms, needed a banquet and had, instead, been thrown a dry sandwich.

The Council condemned the Review as totally unacceptable. Glyngwyn Roberts added fuel to the fire in a powerful speech in which he made a strong attack on the Minister. He likened the industry's position to the fate of some crabs he once saw boiling in a pot. When he asked whether the practice was cruel he was told "Oh, we don't put them in boiling water – we put them in cold water and bring it to the boil!" The Price Review system treated farmers in the same way. He argued that the Union should not give a platform to a Minister who had treated the industry in such a damaging and contemptible manner and he urged the Union to withdraw its invitation to the Minister to address the annual meeting.

I had the temerity to suggest that such action would be damaging for the Union's chances of recognition but the outcome was inevitable – the Council passed a vote of 'no confidence' in the Minister, called for his resignation and agreed to withdraw the invitation to address the annual meeting. It was further agreed that the invitation be extended to Gwynfor Evans MP, the leader of Plaid Cyrnru.

The decision assuaged the anger and militancy of Council delegates but it sent shock waves through the civil service and infuriated other Ministers. A short time later, at a meeting with another government Minister, Union leaders were warned, in no uncertain terms, that all government doors would be closed to it if such intemperate action was repeated. As it turned out, the Union had to wait another eight years before it achieved the all-important goal of recognition. Cledwyn Hughes was not invited back to address the annual meeting until 1981 when, as Lord Cledwyn of Penrhos, he was presented with a memento for his outstanding services to Wales and agriculture.

The presentation went some way to compensate for the Union's actions some ten years earlier. In retrospect, Lord Cledwyn was not to blame for the government's failure to appreciate the needs of the industry in 1970. He had little input into the Price Review having inherited most of the package from his predecessor, Fred Peart. The Wilson government was, at the time, pursuing a cheap food policy as part of its determined commitment to counter inflation. Lord Cledwyn was sympathetic to the industry's needs – he preached the gospel of encouraging home food production in order to reduce imports – a policy which the 'Little Neddy' report of 1968 estimated could save, the UK £200 million a year in food imports.

Lord Cledwyn's proposed 1970 Price Review package was strongly opposed by the Chancellor, Roy Jenkins, who had formidable support from Tony Crossland and Barbara Castle in the Cabinet. They were backed by the Prime Minister, Harold Wilson and Lord Cledwyn had no alternative but to settle for less than he wanted.

Lord Cledwyn was instrumental, in 1965, in arranging the first meeting of Union representatives with Ministry of Agriculture officials and, when Minister of Agriculture, he made a point of meeting Union leaders at the FUW stand at the Royal Welsh Show. When Minister of Agriculture he also decided, against the wishes of his civil servants, to share agricultural responsibilities for Wales with the Secretary of State. This decision allowed the FUW to become more effectively involved in policy and Price Review discussions and led, eventually and inevitably to full recognition of the Union. These decisions, which were opposed by some leading members of the Labour party in Wales, gave an important impetus to the devolution process. In 1977 Lord Cledwyn readily agreed to chair a meeting at the NFU's London headquarters and act as mediator in another vain attempt at unity.

Cash Crisis

The mid 1960s proved to be a very difficult period for the Union – it found itself in severe financial difficulties and there were fears that it faced bankruptcy. The Union's official insurers – National Employers Mutual (NEM) – expressed serious concern about the situation. The insurance company – which played a vital role in the Union's income – feared a heavy financial loss if the Union collapsed and urged the introduction of new, tough measures to control spending. It was particularly critical of the committee system which controlled financial administration. It's view was that it placed the responsibility in too many hands and, as a result, failed to take swift and effective action to achieve objectives. NEM indicated its willingness to fund the Union, but only if the Union could raise a similar sum as a reserve fund which could be used as collateral.

In the following weeks Myrddin Evans and Emlyn Thomas – the Union's new General Secretary – visited every county branch in their quest for donations. The response was startling and exceeded their expectations with individual members subscribing hundreds of pounds. The staff was warned that they may not be paid unless targets were met and economies were introduced..

I had personal experience of the economies – within days my responsibilities were extended to cover the secretaryship of the Union's Policy Committee, and co-administrator of Welsh Farmers Limited – a bulk buying and discount service for FUW members which supplied anything from a TV set to animal feed and veterinary products. At that time Editorship of the Union's journal 'Y Tir' was already part of my responsibility while a Cardiff PR company was responsible for procuring advertising – the lifeblood of the journal. Within a few weeks their services were dispensed with and advertising added to my care.

Within a year of becoming President, Myrddin Evans was able to announce that the position had improved and that the crisis was over – provided there was continued financial discipline. Despite the best of intentions however, financial instability continued to haunt the Union during the next decade.

One of the reasons for the Union's financial problems was the additional burden which it assumed as it attempted to meet its increasing representational role on behalf of farmers and justify its claim for recognition. Administrative costs increased significantly during the 1960s and 1970s, a period when the Union was fully engaged in a number of important issues, not least its opposition to the proposed Rural Development Board for Mid Wales.

Emlyn Thomas – formerly the FUW's Carmarthenshire County Secretary had no formal training in the law – many believed that had he done so he would probably have been a very successful country solicitor or even achieved fame at the bar. A tall, commanding figure he had built up an impressive reputation in handling and advising his Carmarthenshire members on a variety of issues relating to agricultural law. He had been prominently involved in the successful opposition to the drowning of the Gwendraeth valley. He had a sound knowledge of agricultural policy matters and while at Carmarthen he was secretary of the South Wales branch of the National Milk Producers Association which represented hundreds of dairy farmers in the area.

He was however, not a desk-bound administrator, preferring to be out on farms with members or at meetings with Ministry and other officials. He was in his element addressing meetings and the Rural Development Board (RDB) inquiry at Aberystwyth in 1968 provided him with a stage which gave him ample scope to display his oratory and legal skills. When he addressed an audience he adopted an exaggerated, rhetorical pose which endeared him to his supporters and infuriated his opponents. I can recall the sight of a short, slightly built Ministry official hopping with rage and prodding the towering Emlyn Thomas in the chest in an attempt to curtail his criticisms of the Rural Development Board.

Emlyn Thomas represented 1,600 objectors at the RDB inquiry and had a very formidable opponent in W.L. Mars Jones QC who represented the Ministry of Agriculture . Emlyn Thomas was far from over-awed by the occasion and proved to be an outstanding advocate on behalf of those who opposed the Board. The inquiry placed an enormous administrative burden on the Union and its resources – no fewer than 2028 proofs of evidence were prepared for individual objectors – the transcribed documents were released at 7 pm every evening and had to be carefully checked by the following day.

The inquiry was adjourned after three months and re-opened in the autumn. Emlyn Thomas made his final submission on 15th October 1968 – he accused the Ministry of misleading the inquiry and maintained his attack on the Board's compulsory powers to amalgamate small farms. Although the government decided to go ahead with the proposal, the FUW ,with support from the CLA, petitioned Parliament and succeeded in delaying its implementation until the Conservatives won the general election in 1970 and kept its promise not to go ahead with the Board.

Emlyn Thomas found it difficult to get away from an interesting discussion and, as a result, he was sometimes late for meetings. His time-keeping was not improved by a slipped disk which sometimes gave him considerable pain. I invariably accompanied him to meetings and I recall one occasion in Brecon when, before a meeting, he had so much pain that it was doubtful whether he would be able to attend. A local Union official suggested that he visit a farmer in the locality who had a reputation as a healer. Nothing ventured – nothing gained – when we got to the farm we found the farmer waiting for us outside his surgery – an old, dilapidated railway carriage.

Emlyn Thomas told me later that as soon as he entered the darkness of the carriage he was, without warning, gripped from behind and manhandled on to a make-shift platform made of sacks of wool and straw. His back was pummelled and kneaded for some five minutes as the local FUW official and I waited anxiously outside listening to what sounded like a wrestling match. To our astonishment the 'patient' emerged feeling much better.

Many of the meetings we attended were held in public houses and, in West Wales, the problems in the milk industry always drew a large crowd. At the first of these meetings I attended Emlyn Thomas asked me to get him a tumbler of milk to ease his throat and added firmly "Don't forget to say its for me". The barman duly filled the tumbler half full of milk and topped it up with a generous amount of whisky. When the tumbler was placed before him he would clear his throat theatrically and drink the milk before the approving gaze of local dairy farmers.

Emlyn Thomas spent only four years at Aberystwyth, before he left to join the staff of the Liberal party. In 1968 he was succeeded by **Evan Lewis** who had been the Union's Assistant General Secretary for six years before leaving to join the staff of the local authority in 1965. A former RAF pilot he had been a civil servant in one of the emerging African countries and was conversant with tribal differences — a quality which may have given him useful experience for the demands made upon him later in his career!

Evan Lewis was, to paraphrase Gilbert and Sullivan, the very model of a modern civil servant. An outstanding administrator he would not be deviated from ensuring that the Union's work was established on a solid, financial foundation. During his early years at the Union – a period of rapid expansion and escalating costs – he and Myrddin Evans steered the Union through some very difficult periods of financial instability. He later recalled that his initial priority was to ensure that the Union had the financial framework to support its work and activities. He recognised the vital importance of the Union's insurers and fostered an amicable and efficient working relationship with the Union's insurance company.

During one financial crisis, county branches and headquarters staff were given an explicit warning that unless they played their part in controlling expenditure the Union would be finished within a short period. Expenditure at all levels was cut to the bone and the Union survived the crisis.

Myrddin Evans a former publican – and Evan Lewis formed a formidable partnership – Myrddin Evans had the courage to take – and justify – difficult decisions and he proved to be a very effective leader during his 18-year tenure of the Presidency. His early chairmanship of the Finance Committee had given him an unrivalled knowledge and understanding of the Union's finances and he coupled this with a sound knowledge of farming issues which impressed civil servants and Ministers. Although he pursued Union objectives with determination and vigour he invariably gained the respect of his adversaries. He bitterly regretted the fact that he failed to find common ground with his opposite numbers in the NFU in Wales. He established a working relationship with NFU President, Sir Henry Plumb in 1978 and believed that had unity talks been left to Sir Henry and himself they would have stood a good chance of success.

To everyone's surprise, Sir Henry accepted an invitation from Myrddin Evans, to address the Union's annual meeting in 1982 when he was leader of the European Democratic Group in the European Parliament. Sir Henry overcame all obstacles in his determination to honour his promise by chartering a private plane to fly him to Aberporth and fly him back to Heathrow in order to catch a plane to the United States where he was due to meet President Reagan.

Myrddin Evans' failure to achieve an honourable settlement with the NFU was his biggest disappointment during his 18 years of outstanding leadership. He commented later "I would have been immensely proud if we had settled our differences – I would willingly have given up the Presidency for the sake of unity."

Looking back over a period of abortive unity talks he added: "Nobody has worked harder to achieve unity with honour than I have during the last few years; it is heart-breaking to realise that we are no nearer to reaching that objective than we were when I first took office. I believe, quite simply, that this is a tragedy for Welsh agriculture."

Myrddin Evans stepped down from office in 1984 after 18 years of unprecedented progress in establishing the Union as a potent force in Welsh farming. He and Evan Lewis overcame the financial instability of the 1960s and 1970s and provided a solid platform for future expansion. He received the OBE in 1977 for his services to the industry and later the CBE. He was inducted into the Gorsedd at the Wrexham National Eisteddfod and also made a fellow of the Royal Agricultural Society. He later served on the committee of

Carmarthen Farmers Ltd., and as a magistrate on the local bench when he was also a member of the local police authority.

Myrddin Evans was succeeded by Anglesey farmer **Huw Hughes**, another who was better known by his initials, H.R.M. He entered the industry as a farm worker in Carmarthenshire before moving back to his native Anglesey with his young family and purchasing the 25 acre smallholding, Penrhos, near Bodedern, where he was born. He started his dairy herd with three Ayrshires because they were cheaper than Friesians and over the years he extended the farm and established a prize-winning Ayrshire herd of over a hundred. He established a reputation as a noted Ayrshire breeder and won many awards at national agricultural shows and Society competitions. By the time he took over the Presidency he had established a flock of 200 ewes and a 5,000 bird poultry unit producing eggs which, with milk, were sold locally.

In 1958 he became a member of Anglesey's Executive Committee and was elected to serve as one of Anglesey's representatives on the North Wales regional committee of the Milk Board. In 1966 he entered the Presidential ranks and took over as Deputy President in 1970. He served as Deputy President to Myrddin Evans for 14 years and was a loyal and sincere ally throughout that period.

H.R.M. Hughes has strong principles and although he is a quiet, unassuming character he is not afraid to speak openly and honestly on controversial issues when occasion demands.

In an article in the Farmers Weekly, its Wales correspondent, Robert Davies, stated that it would be difficult to find two more contrasting personalities than Myrddin Evans and his successor, "While the Evans style of leadership was laconic and unflappable, Mr Hughes has a justifiable reputation for emotional rhetoric and passionate commitment to causes, a characteristic which often leaves him lost for words, even in his first language of Welsh..." He added that while some members might regret the passing of urbane diplomacy from the old FUW leader, nobody questioned H.R.M.'s absolute honesty and total dedication to the Union.

H.R.M. was a lay preacher and like all good preachers he would – in his early years – become so committed and animated in getting his views across that he would sometimes go into a 'hwyl' (fervour) which could obscure his basic message. This understandable, and some would say, admirable trait, nevertheless posed a problem.

For some 30 years during my period with the Union I was involved in the organisation and orchestration of the Union's most important domestic event – its annual general meeting. This entailed arrangements for the agricultural and national Press – which invariably included TV coverage in Welsh and English. More importantly perhaps, it also entailed writing the President's annual address which followed a familiar format – an introduction referring to the past year's problems and highlights, strong and controversial views on developments in the industry and prospects for the future and, finally, a rousing, rallying call in defence of farming generally and Welsh farmers in particular!

This system usually worked satisfactorily but during H.R.M.'s early Presidency the content of the 'script' was sometimes threatened by his zeal and enthusiasm. As a result, in order to combat this tendency, I positioned myself carefully directly in his field of vision, behind the seated delegates and TV cameras, and shook my head vigorously or even waved surreptitiously to stem the flood of words from the rostrum. H.R.M. never objected to this treatment and he developed into an effective speaker – and a brilliant interviewee on radio and TV – particularly in the Welsh language.

H.R.M. Hughes was President of the Union when Secretary of State, Peter Walker encouraged unity talks in 1987. The following September the NFU indicated that it was prepared to discuss a merger with the FUW but it took another year before H.R.M. and Simon Gourlay, the NFU President, found themselves across the table in Cardiff with Welsh Office Secretary John Davies in the chair. As H.R.M. feared, the talks subsequently ended in failure. Nevertheless, he emphasised his regret that it had not been possible to unify the two organisations.

H.R.M. Hughes grew in stature during his Presidency and when he finally stepped down to make way for another Anglesey farmer, he could look back with satisfaction on his period of leadership.

Bob Parry, his successor, a beef and sheep producer who farms at Bryngwran, near Holyhead, had an ideal background having served for some years as Chairman of the Union's Livestock Committee. He was a member of the Anglesey Council and Mayor before local government reorganisation when he continued as a member of the island's County Council and was elected its leader. He had also been chairman of the Council's planning committee. He was a member of the Meat and Livestock Commission's Liaison Committee and of its Beef Promotion Council. He received an OBE for services to agriculture in 1994.

Bob Parry's Presidency – from 1991 to 2003 – coincided with a period when farming went through a series of crises which placed additional demands on the Union and an almost intolerable, physical burden on its leaders and livestock representatives. The BSE crisis, the ban on livestock exports, allied to the pressures exerted by animal welfare activists, the foot and mouth outbreak of 2001 and the early stages of the reform of the European Union's agricultural policy, were all of direct concern and importance to Welsh farmers and the Union.

The establishment of the Welsh Assembly, which was warmly welcomed by the Union, opened more doors for representation and, for a leader based in Anglesey, there was always the initial trek to Union headquarters at Aberystwyth, and then the extra hundred or so miles to Cardiff before getting down to the agenda and the work in hand. Bob Parry's work-rate was prodigious and he maintained this demanding pace to earn a reputation as one of the Union's most industrious leaders. The 'politicking' local authority syndrome, some considered, sometimes affected his judgement but he led the Union very effectively at a period when lesser individuals would have capitulated to the pressure.

The foot and mouth outbreak with its catastrophic impact on farming and the rural economy generally – was the most traumatic blow for farmers during this period. Many saw their cattle and sheep, painstakingly bred and developed over the years, perish on open, funeral pyres. A total of six million animals were slaughtered in the UK and the cost to the economy was over £3 billion. Around

350,000 livestock were slaughtered in Wales before the outbreak was brought under control.

Bob Parry described the period as 'truly horrific' while Livestock chairman Alan Gardner said it had been the most difficult year in living memory.

The Union formed a close working relationship with all the authorities involved in countering the outbreak but had reservations about the effectiveness of the cooperation between government agencies. Whilst the Department of the Environment, Farming and Rural Affairs (DEFRA) had overall responsibility for eradicating the foot and mouth outbreak in the UK, the National Assembly had responsibilities for only certain aspects of disease control and lacked the primary legislative powers necessary to react with appropriate urgency.

The Union complained that a common factor underlying all disease outbreaks was the need for much stricter border controls to prevent the importation of infective agents. It continues to press for tighter controls at points of entry to the UK and has succeeded in raising public awareness of the serious threat to animal health, and the economy.

Bob Parry later remarked that the best thing to come out of the FMD outbreak was the cooperation between farming organisations. He commented: "It proves that we can work together for the benefit of the industry."

Welsh Assembly

Bob Parry's Presidency marked the achievement of one of the Union's most important objectives – a Welsh Assembly with sweeping powers over Welsh agriculture. The Union had wholeheartedly supported the proposals outlined in the government's White Paper, 'A Voice for Wales' which proposed the transfer of most of the powers of the Secretary of State to a Welsh Assembly. The proposals were endorsed by a referendum in 1998 and the Assembly was officially opened a year later

It was ironic – in view of the Union's enthusiastic support for the Assembly – that the Assembly's first Agriculture and Rural Affairs Minister was Christine Gwyther, a vegetarian! The appointment infuriated the farming community and Bob Parry walked out of his first meeting with the Minister as a protest commenting "Here we are on the verge of launching a new Welsh food strategy at a time when we want more people to buy and eat Welsh beef and lamb. And who does Alun Michael put in charge of it all – a vegetarian who, as a point of principle, has decided never to eat meat!"

The Union subsequently warmly welcomed First Minister Rhodri Morgan's subsequent reshuffle which brought Carwyn Jones to the rural Minister's post.

Bob Parry was an energetic leader who committed himself totally to the Union. His commitment and work rate during the BSE crisis, the foot and mouth outbreak and other problems which beset the industry at that time can only be described as extraordinary He made the most of his sound knowledge of the livestock sector and was an effective communicator, particularly in the Welsh language. His sometimes fiery temperament added conviction and sincerity to his media interviews and speeches but his assertive manner could, at times, be to his disadvantage in the ultra-democratic environment of the FUW.

Bob Parry was succeeded by the Union's first President from Montgomeryshire, **Gareth Vaughan** who took over the reins in 2003.

NEM & Insurance

The FUW's insurance service has played a vital role in the stability and development of the Union. One of the Union's early priorities was to offer its members the same range of insurance services as its competitors. The commission it has earned has contributed very significantly to the Union's coffers. The Union's link with its initial insurers, National Employers Mutual (NEM) was forged by the NEM's Cardiff manager who heard about the Union's establishment and offered the company's services to the new Union. Following initial talks between the leaders of the Union and company representatives, including company manager G. Percival, the Union had no hesitation in coming to an agreement with NEM. It is doubtful whether the Union would have survived some of the financial pressures of the early years without the support of the NEM. Apart from commission the company financed its presence at major agricultural shows and supplied the Union with rented premises it purchased in key areas of Wales.

One of the reasons for the success of the link between the two organisations was the decision to transfer Tom Astley, a native of Llanuwchllyn, from London to Aberystwyth to serve as Liaison Officer between the two organisations. Tom Astley proved to be a popular figure with Union leaders and staff and was an outstanding success in settling the teething problems which arose in the early years. He was sympathetic to the Union's aims and played a vital role in assisting the Union when it ran into financial difficulties in the 1960s. Tom Astley retired in 1973 and was succeeded by Evan Evans, who had risen from the 'ranks' and proved equally effective and popular.

The NEM expanded its services to life assurance through National Employers Life (NEL), a service which was already catered for by the Scottish Provident Institution (SPI). Over the years the changes in the insurance market has seen the amalgamation of various companies and the Union's insurance services, which continue to be of vital importance, have progressed from NEM to AGF, to Cornhill and the present Allianz/Cornhill.

Political Independence

The Union has always been jealous of its political independence and has been ultra careful not to support or align itself with any political party or movement. The political views of the leading founder members of the Union were so well known that even in its early days its critics hesitated to justify a political motive for the Union's establishment. Nevertheless, the increasing interest in political devolution raised the issue of nationalism – in reality however, only one of the Union's 14 founder members – Llew Bebb of Goginan – had known links with the nationalist cause.

The three prime movers, J.B. Evans, Ivor Davies and D.J. Davies had contrasting political beliefs – the former was a well-known Tory and contested Carmarthenshire as a Parliamentary candidate while the latter had close links with the Labour party in Cardiganshire and also contested the county in a Parliamentary election. Ivor Davies was a prominent Liberal. The founder members regarded political independence as crucial and Ivor Davies made this clear from the commencement "The Union is non-political and must always be so. To serve the industry fully it must always be free to criticise, if necessary, the views of all political parties."

When the Union selected National Employers Mutual as its official insurers in the weeks after its establishment it gave the company a clear undertaking that the FUW was a non-political organisation and would stick to this commitment. In 1959 a Plaid Cymru candidate in North Wales was told that he could not be considered for the post of Assistant General Secretary and Legal Adviser with the Union because he was not willing to give an undertaking that he would not take an active part in party politics after the election. It was pointed out that two previous members of staff, J.B. Evans and John Morris QC, later Secretary of State for Wales, had resigned from their posts with the Union to contest Parliamentary seats.

In an article on the Union, not long after its establishment, the Financial Times said it would be a mistake to attribute the Union's establishment to Welsh nationalism. It was, it stated, more the result of the frustration and dissatisfaction relating to farming's problems.

Another political activist who had close and early links with the Union and played a prominent part in the Union's recruiting campaign in Mid and North Wales was a young Geraint Howells from Ponterwyd, Cardiganshire's first county chairman, later chairman of the Union's Livestock Committee, who won the Cardiganshire Parliamentary seat for the Liberals and was later elevated to the House of Lords as Lord Geraint.

Ironically, the only time I was required to defend the Union's political independence was in 1976 following the publication of the Scottish NFU's report on its organisation and future structure. The report emphasised the SNFU's determination to retain its independence within a federal system of UK representation – exactly the role the FUW sought.

The SNFU report stated: "We begin this report by recording our emphatic support for the simple proposition that the SNFU must remain an essentially independent national organisation while, at the same time, maintaining effective working relationships with other organisations, and in particular with the NFU of England and Wales and the Ulster FU... We believe that our independence has been to the advantage of the Scottish farming industry in the past and we are convinced that this will continue to be so in the years ahead."

In the ensuing publicity which accompanied this report there was a reference in the 'Scottish Farmer' to the FUW's "strong nationalist element." Following the publication of my letter emphasising the FUW's political independence and the striking similarity between the SNFU and FUW's views on a federal structure of representation I received a number of letters from Scotland supporting the FUW's policy.

The Union's determination to maintain its political independence was emphasised by FUW President H.R.M. Hughes after he had been described as having "a strong Welsh background". H.R.M. nailed his colours to the mast when, within a few weeks of taking office, he stated "I do not support the FUW because of some feeling of nationalism. Farming and the FUW are bigger than party politics and must remain so. We have to work with whatever party is in power."

Early Days

A 1967 meeting at Aberaeron with Francis Pym MP, and Nicholas Edwards MP. Mr Pym gave an assurance that there might have to be some changes to enable the Milk Marketing Board to comply with the Treaty of Rome but that its "essential marketing functions" would continue!

An early picture of county and headquarters staff includes Emlyn Thomas, Glyngwyn Roberts, Emrys Benett Owen, R.P. Davies, Meurig Voyle, Walter Rowlands, R.O. Roberts and R.J. Williams.

He'll soon come back to mother!

The FUW was a fruitful source of material for cartoonists but one of the first – by JC Walker in the Western Mail shortly after the Union's establishment – predicted a short life for the Union. However, it made the Union's early leaders even more determined to survive. Within a few years the Union was said to be celebrating financial success – a cartoon in the British Farmer and Stockbreeder by Chrys (below) showed Union officials counting record subscriptions. Unfortunately it did not last and within a few years the Union was struggling to survive a financial crisis.

"NEVER IN THE FIELD OF 'UMAN ENDEAVOUR HAS THE FUW OWED SO MUCH TO SO MANY!"

The first FUW meeting with a Minister of Agriculture – Fred Peart, accompanied by the Ceredigion MP at the time, Elystan Morgan, with FUW leaders R.O. Hughes, John Evans, H.R.M. Hughes, Myrddin Evans, D.J. Davies and Gwilym Thomas.

Liberal leader David Steel, now Lord Steel at an early annual meeting.

Cardiff rally in 1990 with Lord Elis Thomas.

Former head of ADAS and Principal of the Royal Agricultural College, Sir Emrys Jones at an Aberystwyth annual meeting of the Union.

EVOLA
EVOLA

Above: The current headquarters at Gogerddan, near Aberystwyth. Opposite (top): The Union's Chalybeate Street headquarters in Aberystwyth from 1956 to 1963. Opposite (bottom): Queen's Square headquarters in Aberystwyth from 1963 to 1999. Opposite (inset): First FUW headquarters 1955/56, Gwalia House, Carmarthen.

A Pride of Presidents – Farm Minister Nick Brown who opened the new headquarters in 1999 with (left to right) former Presidents the late Glyngwyn Roberts, Myrddin Evans, H.R.M. Hughes and President Bob Parry.

Influential Pioneers: J.B. Evans (left) with Glyngwyn Roberts and Geraint Howells – later Lord Geraint – were, with D.J. Davies and Ifor Davies, the most influential of the early members of the FUW. Glyngwyn Roberts expressed the views of North Wales members during the stormy early days. Geraint Howells played an important role in his native Ceredigion and in gaining recognition. J.B. Evans steered the Union through its difficult early days and described it as a baptism of fire.
Below, Gwilym Thomas pictured with Evan Lewis in Council – Evan Lewis and Myrddin Evans were instrumental in placing the Union on a sound financial footing.

Secretary of State David Hunt with FUW President Bob Parry and the late Alan Lewis.

Milk Board member Wallace Day with FUW President Myrddin Evans and Denbigh representatives, including Llysfasi Principal Maldwyn Fisher.

The First FUW

An earlier attempt to establish a Welsh farming Union in 1918, collapsed four year later and the organisation was absorbed by the NFU. The Union – its official title was Undeb Cenedlaethol Amaethwyr Cymru (National Farmers Union of Wales) was established at Rhyl on the 8th November 1918 and its aims bore a striking resemblance to those of the FUW.

The President of the Union was Mr J.N. Thomas from Trefignath, Anglesey and the two Vice Presidents were J. Jones of Berth Ddu, Llanrwst and D.S. Davies of Ty Coch, Ystalyfera, near Brecon. The General Secretary was a Mr John R. Chambers whose office, presumably the headquarters, was in Llanrwst. Included in the Union's 13 aims was the general advancement of Welsh agriculture and the interests of Welsh farmers, to pursue test cases for Welsh farmers, assist in the promotion of Welsh cooperatives, cooperation with other Unions and agricultural organisations, the development of transport facilities in Wales, improved insurance terms and support for agricultural education.

Little is known of the Union's progress in the four years after its establishment but it was sufficiently strong to merit a resolution from the Council of the NFU which proposed the amalgamation of the two Unions. Although the terms of the amalgamation gave the impression that the Welsh Union retained its independence, on closer examination control passed entirely to the NFU. All income was remitted from Wales to the NFU head office and the NFU kept tight control of expenditure.

A year after the FUW was established – in 1956 – the late Tom Williams of Forden, Montgomery, a former President of the NFU, said that the time was not opportune to form a separate Union for the farmers of Wales. He added that such a Union should be established when Wales had its own Secretary of State. Mr Williams died just a week before the government named Mr James Griffiths, MP for Llanelli, as the first Secretary of State for Wales.

County Stalwarts

During my period with the FUW I was fortunate to get to know and work with several respected and popular County Secretaries. A good example was **Walter Rowlands** of Glamorgan who played a leading role in the Union's development in the county. He was involved in the Union from its establishment and was a member of staff for nearly 30 years. He had experience of farming's problems and over the years he came to represent the views of many who were neighbours of large industrial units and valley housing estates and experienced problems rarely encountered by farmers in more isolated, rural areas. Walter Rowlands played a major role in solving sheep straying problems in the South Wales valleys by campaigning for the installation of cattle grids. He was instrumental in establishing the pioneering Ogmore and Garw UDC scheme to prevent sheep straying which provided the blueprint for many similar schemes in later years. He was succeeded by the equally long-serving **Angela Giddings**.

Anglesey was fortunate to secure the services of **R.J. Williams** in 1959, a former solicitor's clerk. He built up a huge personal following on the island and rejected an offer to join the NFU as their County Secretary. He had an outstanding record for servicing members and one of his most notable achievements was his success in gaining significant increases in wayleave payments for his members who were affected by Shell UK's pipeline through the county after other organisations had accepted lower payments. He also played an important role in attracting Halal Meat Packers to Gaerwen.

The Anglesey County Executive committee was also fortunate to have the Minister of Agriculture, Cledwyn Hughes (later Lord Cledwyn) as their Member of Parliament for several years and benefitted from his numerous attendances at their meetings. R.J. was awarded the MBE for his services to the agricultural industry in 1984 but died at the relatively early age of 64. Michael Dolan succeeded R.J. Williams.

Another long serving member of staff was **Owen Slaymaker** in Carmarthenshire, a former colonial officer who, along with Esme Lloyd, a secretarial officer, established an outstanding record of service to its large number of members. After Owen Slaymaker's untimely death he was succeeded by **Peter Davies** [now the Union's Director of Administration]. Carmarthenshire has had a large number of outstanding field staff – one colourful character fondly remembered by a large number of his members was **Horace Miles**, an ex-Army physical training instructor who looked after his members with boundless energy and good humour – and always gave the impression that he had just stepped off the parade ground!

Another colourful character to emerge from Carmarthenshire was **Meurig Voyle**, who remembers the tempestuous early meetings of the Union in West Wales. He joined the Union's staff as Assistant Secretary in 1961 and worked with County Secretary Emlyn Thomas who took part in the successful campaign to prevent the drowning of the Gwendraeth valley and later moved to Aberystwyth as the Union's General Secretary. In 1966 Meurig Voyle moved to the Denbigh office of the Union on a temporary basis to organise the Union's North Wales counties during the regional Milk Board election involving FUW candidate Glyngwyn Roberts and J.M. Evans which the former FUW President lost by only 800 votes.

Meurig Voyle was a veteran of the second World War and had seen active service with the Royal Artillery and had been awarded Field Marshall Montgomery's Certificate. Unfortunately for him the county office at the time was in the Old Back Row Hotel in Denbigh – old being the operative word as it had no heating and few amenities. Meurig Voyle slept in the office during the election campaign and swore that, despite his military service, he had never suffered such privation! This introduction to life in North Wales did not influence his decision when, a little later, he accepted the post of County Secretary in Denbighshire and moved with his young family to North Wales. Meurig Voyle spent over 30 years servicing the needs of FUW members in North and South Wales and later recorded another 15 years in the service of the Welsh Mountain Sheep Society. He was made a life member of the Union in 2005.

Clwyd's Cold War

Meurig Voyle can tell his grandchildren and great-grandchildren that he played a part in ending the East-West cold war of the last century. He hit the headlines of the national Press in 1980 when a leading member of the Union, Aubrey Morris, walked into the county office carrying a radio transmitter which had been unearthed on his farm by his son when he was ploughing.

Meurig Voyle was not one to pass up an opportunity of Press coverage and he notified national and local newspapers which carried pictures and reports about the mysterious find. The transmitter was taken to the police but when they were later asked where it had come from he came up against a wall of silence. Later a question in Parliament elicited that the transmitter was of Soviet bloc origin.

Four years later all was revealed when KGB defector Stanislav Levchenko admitted that the transmitter had been used by Russian agents spying on nuclear submarine movements in the Merseyside area. It was common knowledge that a number of Russians, said to be part of a trade mission, had stayed in North Wales, including a group at the Wynnstay Hotel at Llanrhaeadr-ym-Mochnant. Five members of the 'trade mission' were later deported for 'unacceptable activities'.

Levchenko was regarded as one of the top agents in the Russian KGB and a major catch for the Americans. He appeared on TV in the United States and admitted that he had been a member of a UK spy ring monitoring nuclear submarine movements and that the information had been relayed back to Moscow by radio transmitters. It is believed that the Clwyd 'trade mission' had abandoned and hidden the transmitter when UK security services moved in on the group.

Meurig Voyle was one of a number of FUW county staff who brought their own distinctive style to the important servicing and advisory work provided from local offices. Whilst local government and other farming organisations rationalised their administrative centres, the FUW stuck stubbornly to its old county structure despite, at times, strong financial pressures.

Other long-serving County and Area Secretaries members will recall are **R.P. Davies** of Montgomeryshire who took up the cudgels on behalf of Mid Wales farmers threatened with new reservoirs, new towns and National Park and the Rural Development Board. Others include J. Dyer James (Merioneth), Brian Edwards (Gwent), Herbert Miles (Pembroke) and Lewis Griffith (Ceredigion). Long serving county staff included Huw Jones (Merioneth), John Roberts (Brecon & Radnor), Neil Smith (Gwent) and the previously mentioned Michael Dolan (Anglesey) and Angela Giddings (Glamorgan) – Lyn Williams and Mansel Charles (Carmarthenshire), Gwyn Williams (Denbigh), Dei Charles Jones (Merioneth), Colin Greaves (Glamorgan & Gwent) and the late Dewi Hughes (Pembroke) and E.D. Jones (Ceredigion).

Ministerial Meetings

Meetings with government Ministers were few and far between in the early days of the Union but as it became more established Union representatives met Welsh Office and Ministry of Agriculture Ministers on a number of occasions. Each meeting raised hopes that the Union would soon be accepted as an officially recognised body representing Welsh agriculture. After recognition, meetings with Ministers were regarded as normal practice and the natural culmination of government representations. After entry to the Common Market, Union representatives had easy access to Agriculture Commissioners and their specialised staff and forged a stronger link with Members of the European Parliament.

In 1996 President Bob Parry attended industry summit meetings with Prime. Minister John Major on BSE and two years later with Prime Minister Tony Blair on the European Union's beef export ban. Although crises meetings of this nature are rare there have been more pleasant meetings with leaders in the past.

In the early days of the Union, Prime Minister James Callaghan visited D.J. Davies' Pantyryrod farm, near Aberaeron to purchase calves for his farm. Glamorgan FUW chairman John Llywellyn attracted considerable publicity when he presented a calf on behalf of the branch to Prime Minister Ted Heath during his visit to Barry in 1974 in order to draw attention to the difficulties in the livestock sector and the collapse of calf prices. Ted Heath took the gesture in good spirits but refused to comment on the calf's eventual fate!

Two years earlier members and guests at the annual dinner of the Carmarthenshire branch of the Union had a surprise when they were joined by Harold Wilson, leader of the Opposition at that time who was staying at the same hotel where the dinner was held, after a tour of West Wales.

Mr Wilson was in sparkling form when he made an impromptu speech after being introduced by President Myrddin Evans. Mr Wilson apologised for turning up in his working clothes – a reference to the large number wearing evening dress. He added that if he had known earlier he would have come dressed as a working farmer! He turned to the Christmas menu and noticed that turkey was the main course. He chided the farmers for not eating Welsh beef – he said that the beef he had eaten at the hotel earlier in the evening was the best he had ever tasted. Mr Wilson, who was not a fan of the Common Market, said that perhaps they had exported all their beef to the Continent! He added that the Tory Government, led by Kent MP Ted Heath had rushed the country into the Common Market before realising that thing's were going to be very different. He wished farmers the best of luck – they would need it!

As far as horticulturists in Kent were concerned, added Mr Wilson, they were in for a nasty time. And with a direct swipe at Mr Heath be said that those in Kent deserved what they got anyway!

Sacre Bleu!

My strangest experience of national leaders occurred not in Britain but in France during the Paris Show of 1973. With FUW President Myrddin Evans I attended a farming function at the George V Hotel, described as the most exclusive hotel in the city. One of the speakers was the French Minister of Agriculture, a tall, thin Jacques Chirac – later President of France. He had already earned a reputation as the champion of French farmers interests. A controversial figure – and speaker – he hit the headlines on more than one occasion for his criticism of the United States, West Germany, UK and EU Commission policies when they conflicted with French interests.

Chirac's speech at the function was not popular with some of his English listeners who grew increasingly restless and hostile. One, sitting opposite me on the same table, who had, I suspect, had too much wine, started to greet every word with a raspberry or profane rejoinder. Those seated around him, including myself, regarded him with some trepidation and inched away as far as possible. Chirac ignored the constant interruption and took it all in his stride, no doubt, expecting nothing better from his neighbours across the Channel!

The Paris Show is a noted shop window for the farm products of all the European nations and Myrddin Evans and I were very surprised that there was a complete absence of Welsh produce. We found that 80 different UK companies and breed societies had publicity stands on display at the show but that there was not one from Wales.

In a statement when we returned, Mr. Evans said that it had been anticipated that in those early days of UK membership of the Common Market, lamb would follow whisky as one of the biggest export products to Europe. He added: "On the evidence of this year's Paris Show it will not be Welsh lamb." He added that it was a tragic situation, particularly as Welsh organisations had the same access to government export funds as others in the UK. He said that he had been impressed by the publicity stands for Aberdeen Angus and Scotch Quality Lamb while the Irish deserved praise for their promotional professionalism. He added that Wales would have to wake up to the opportunities abroad or be left with the crumbs off the table.

The Union pursued the issue in subsequent talks with the Welsh Agricultural Organisation Society (WAOS) and Welsh Quality Lamb Association and the position has improved to such an extent that Welsh agriculture and its promotional agencies can now be proud of their success abroad.

The Union continues to play an active role in Welsh food promotion – in 1987 it welcomed the Meat and Livestock Commission's decision to provide funding for the promotion of Welsh lamb. But the Union had to wait another ten years before the MLC agreed to spend a proportion of its beef promotion levy on Welsh beef. At that time – in 1997 – the Union played a prominent role in a new 'Taste of Wales' poster campaign which was launched at the Royal Welsh Show by the Secretary of State, Ron Davies MP.

In more recent times the Union has taken its campaign a stage further by organising a three-city tour to promote Welsh food during Farmhouse Breakfast Week. The tour, taking in Brussels – where Farm Commissioner Franz Fischler received a hamper of Welsh food from FUW President Gareth Vaughan – Westminster and the Welsh Assembly was so well received that a large group of cross-party MPs tabled an Early Day Motion congratulating the Union on its food promotion efforts.

The Union has also been active in campaigning for the use of more local foods in schools and in other suitable local and central government establishments.

FUW President Gareth Vaughan summed it up when he said "I believe our efforts have reinforced the good work done by other agencies such as the WDA and Hybu Cig Cymru in promoting high-quality Welsh meat and dairy products both at home and abroad."

Royal Support

The Union has a high regard for the Prince of Wales who has generally championed the cause of the farming industry in the UK and Wales in particular, and has not been afraid to express his views on pressures on the industry and the rural economy. My first association with the Prince goes back to 1983 when Ceredigion was the Royal Welsh Show's host county. I acted as publicity officer for the county committee and produced a 100 page brochure – 'Research Into Practice' – on the work of the Welsh Plant Breeding Station – now IGER.

Prince Charles – as Chancellor of the University of Wales – supported Ceredigion's efforts to raise funds for the establishment of an Overseas Pavilion at Llanelwedd – now a reality – and approved a message of good wishes on the Committee's efforts in the brochure.

The Prince attended a special reception for the Union at Powys Castle in 1996 when he was introduced to Union leaders. In August 2003 the Union recognised the Prince's championing of the industry and rural economy when FUW President Gareth Vaughan and Bob Parry presented him with a special award of a suitably inscribed silver salver.

Gareth Vaughan said at the presentation – at a private reception at Vaynor Park, Berriew – "The Prince has proven himself to be a true friend of farming over many years. He has taken a personal interest in the future of the small family farms of Wales and the problems they face."

The Prince and Duchess of Cornwall have, since attended the Union's 50th anniversary reception on Gareth Vaughan's farm at Dolfor, near Newtown.

Brussels Link

The Union had strong reservations about Britain's entry to the Common Market. The issue occupied the Union's attention from the early 1960s usually with varying degrees of scepticism, ranging from cautious acceptance to outright rejection. In 1962 a special, joint meeting of the Agriculture and Policy Committee reviewed months of gathering evidence and finally advocated a number of safeguards and a long transitional period in the event of entry.

The Union continued to consult specialists on the issue and Union representatives travelled to Brussels and other European cities and met a number of Common Market representatives in an attempt to finalise its policy on entry. In 1966 a special Study Group met Georges Rencki, the EU's Liaison Officer with COPA, the Federation of EU farming Unions, to discuss producer representation.

In 1967 the Union accepted an invitation to give evidence to a House of Commons Select Committee on the subject and consulted the Agricultural Economics Department at UCNW Bangor before submitting evidence.

The Union's final submission amounted to a rejection of entry proposals, and based its opposition on doubts about the future of Marketing Boards, the fear that European imports would flood on to the UK market, the possible ending of production grants, the impact on the small, family farms of Wales, particularly hill farms in view of their distance from their main markets, fears about Commonwealth imports and reservations about the willingness of UK governments – of all political colours – to withstand pressures from France and West Germany which gave their farming industries whole-hearted support.

In 1967, FUW President Myrddin Evans described the Common Market as an 'ominous dark cloud on the horizon' and despite a strong recommendation for entry from annual meeting guest speaker

Jeremy Thorpe, leader of the Liberal Party, Union members were strongly influenced by a stirring speech from Union founder member D.J. Davies who warned that rural Wales would be decimated and that empty farms would stand like tombstones on the hillsides of Wales. The outcome was a foregone conclusion in D.J.'s favour.

The Union continued to voice its fears when Common Market negotiations were finalised in 1971 but it realised that entry was inevitable and concentrated its energies on attempting to obtain safeguards for the industry. The Union engaged in a series of talks with the Ministry of Agriculture and General Secretary Evan Lewis travelled to Brussels for meetings with European farming representatives including COPA which represented the European farm unions.

A party of Union members from Carmarthenshire visited farms in Belgium, Germany and the Netherlands and discussed policies and prospects with farming representatives. Evan Lewis and Myrddin Evans continued their talks with COPA representatives but failed, in their attempt to gain independent membership for Welsh farmers. COPA's policy was to treat the UK as one entity as far as membership was concerned and as the NFU already represented the UK and was not willing to accept the FUW there was little hope of membership.

President Myrddin Evans commented: "We have a contribution to make in Europe and Welsh farming deserves independent representation. The tragedy is that we shall not be allowed to make that contribution through COPA."

In 1972 a four man delegation comprising Myrddin Evans, Deputy President H.R.M. Hughes, Policy Advisor J.B. Evans and General Secretary Evan Lewis held a series of meetings in Brussels and were encouraged by the ease of access to EU Commissioners and their policy advisers. In fact the Union discovered that, unlike the Ministry of Agriculture which, at that time, maintained a rather prescribed, formal approach to meetings, similar meetings with Brussels Commissioners and officials were not only easy to arrange, often at short notice – but were also conducted in a friendly, informal manner.

A few years later the Union forged a direct link with Farm Commissioner Finn Gundelach and other EU departments which meant that urgent representations could be channelled directly to Brussels and not through UK government departments.

Myrddin Evans described the direct access as an important breakthrough: "I believe it to be an improvement on COPA's approach as it gives Welsh agriculture a distinctive voice. COPA represents all European farming organisations and they speak with one voice. What suits French or German farmers may not suit us. Certainly I can think of a number of issues on which we disagree with the policies and actions of French farmers."

Over the years the Union has improved its links with the EU and was one of the first organisations to become a member of the Wales European Centre in Brussels. Since the establishment of the Welsh Assembly it has supported, joint representations on Welsh farming and rural issues. In 2002 the Welsh Assembly invested more resources in the Assembly's Brussels office. Announcing the development in May 2002 First Minister Rhodri Morgan justified the expansion by singling out the farm structural funds and CAP reform as issues of huge importance to Wales which were to be finalised in the short term. The Union warmly welcomed the news as evidence of the Assembly's important role in Brussels.

In more recent times the Union has submitted evidence on key issues affecting farmers in Wales – notably the sheepmeat regime while, in February 2003, FUW dairy representative Bryan Jones gave evidence to the European Parliament's Agriculture and Rural Committee on the future of the milk sector. In December of the same year, Deputy President Emyr Jones lobbied the European Parliament, on proposals for the individual tagging of sheep, and union leaders and the Welsh Assembly succeeded in adapting the Single Farm Payment scheme to their advantage.

The Brussels Riot

The Union's success in gaining direct access to Common Market Commissioners and their specialist staff was a situation which needed to be exploited in the constant drive for new members. In 1980 I organised a visit for around 70 farmers to the Brussels farm fair and the Common Market's headquarters in Brussels.

The party was made up of farmers wishing to see at first hand prize-winning Continental livestock and machinery at the Food and Farming Fair – one of the largest in Europe – and Union leaders and commodity delegates who were to meet Common Market specialist staff for a series of pre-arranged policy meetings – and, as many representatives of the media who could be persuaded that the visit was worth covering and reporting on their radio and TV programmes and in their newspapers and trade magazines.

The media response was overwhelming and the visit resulted in a number of reports on radio, coverage on TV, a special half-hour programme on TV and coverage in weekly and daily newspapers and weekly farming journals. The Farmers Weekly, for example, devoted a thousand words and used five large photographs on the visit.

Two MEPs, Ann Clwyd and Beata Brookes helped to arrange the policy meetings on one day in a series of sittings, while Ceredigion MP, Geraint Howells, later Lord Geraint, flew from Heathrow to Brussels to take part in the discussions which centred on that period's burning issues – problems in the beef sector, regional aid and the Less Favoured Areas, the Green Pound, the co-responsibility levy and compensation for farmers who had lost a large number of livestock in a blizzard earlier in the winter.

The three-day visit was not without its problems – on the outward journey one of the 'buses broke down ten miles from Antwerp. Earlier, I had been horrified when the leading 'bus stopped on the hard shoulder of the motorway and disgorged a flood of farmers anxious to respond to the call of nature – some of whom even crossed the central reservation in busy traffic to the other side of the road to do so!

The series of meetings scheduled for one day at the Common Market headquarters on Rue de la Loi coincided with a demonstration by Belgian steelworkers which brought traffic to a crawl as we neared the centre of the city. At one stage our coach was hemmed in by stationary vehicles. The only escape route was blocked by a parked taxi whose driver stubbornly refused to move on to the pavement to allow us through. My pleas fell on deaf ears but when I was joined by half-a-dozen burly farmers willing to lift the taxi on to the pavement the driver cooperated!

When we arrived at the Common Market's Berlaymont headquarters we could see the reason for the delay – the building was surrounded by thousands of steelworkers – some of whom had breached the ranks of police and security personnel and made their way to the top of the building and unfurled a long scarlet banner which was greeted with cheers from their colleagues below.

The FUW party found its entry barred by a helmeted force of surly riot police armed with batons and what looked like sten guns. Fortunately MEPs Ann Clwyd and Beata Brookes who were waiting for us inside the building, managed to convince the police and security staff that we were harmless and allowed us in. We were also fortunate to have been met outside the building by senior EU official Aneurin Rhys Hughes who succeeded in pacifying some steelworkers who felt we were being given preferential treatment. Once inside the building the FUW delegates, led by Myrddin Evans, and his deputy H.R.M. Hughes, embarked on a series of meetings with a number of Commissioners and their staff. The accompanying members of the Press, TV cameras etc, were allowed to remain during the meetings.

Myrddin Evans said later that it was the worst day he had experienced as the Union's President. "I had to deal not only with the Common Market officials but also do interviews with radio, television and newspaper reporters. I felt that the responsibilitv for Welsh agriculture lay heavily on my shoulders and that the eyes of the world were focused on me."

Myrddin Evans and his team did an outstanding job during that day and the publicity it generated exceeded all expectations. Farmers back in Wales saw Union delegates raising relevant issues with the decision makers in Brussels. The visit was a phenomenal success which prompted lavish praise from General secretary Evan Lewis – "The stage having been so well set by you, it was a matter of great pride to see Union spokesmen, in particular the President, rising magnificently to the occasion and giving performances which pinpointed the fact that Welsh farmers interests were being represented directly and forcibly – and by implication – that this had not been done before by other means or groups."

FUW delegates who attended the Berlaymont meetings that day looked back with some amusement at our involvement in the steelworkers demonstration and confrontation with the riot police – tending to equate our helmeted and gun-toting sentinels with the same sense of justice and good humour as the village bobby back home. They added a touch of colour and the opportunity for good-humoured banter – it was not until we returned to the hotel and saw the Belgian newspapers the next day that we realised the full extent of the riot which had brought Brussels to a standstill.

The local TV channels showed pictures of 7,000 steelworkers from Liege and Charleroi 'armed' with banners, marching on the Belgian Parliament and the Berlaymont building demanding government and EEC aid to restructure the steel industry. The steelworkers refused to disperse and started to throw paving stones, flower boxes, and fire crackers at the mounted police. The riot police charged the demonstrators after tear gas had failed to break up the crowds of protestors in front of barbed wire fences.

Subsequent Press reports indicated that 16 riot police were injured, two seriously after being thrown from their horses during collisions with parked cars. Four horses were injured in attempting to hurdle parked vehicles and in clashes with the demonstrators. Those of us who saw the TV coverage shook our heads in disbelief as we appreciated how close we had come to the mayhem in the streets below us.

FUW representatives, led by President Bob Parry, walked into another riot nearly 20 years later when the Union delegation attended Agenda 2000 proposals to reform the Common Agricultural Policy. On this occasion the demonstrators were European farmers who clashed with riot police.

Some years earlier I accompanied FUW President Myrddin Evans and North Wales MEP Beata Brookes to a meeting with the Agriculture Commissioner, Finn Gundelach and his chief adviser, Erik Petersen. Thirty minutes before the meeting we were told that the Commissioner was tied up in some urgent discussions and would not be able to meet us. Mr Petersen would, we were told, deputise for Mr. Gundelach. Both Myrddin Evans I were prepared to accept this but not Beata Brookes who told Mr. Petersen in no uncertain terms that this was a slight on the President of the FUW! After much wrangling Mr Gundelach gave in and abandoned his meeting for a detailed discussion on our representations. Beata Brookes was a formidable advocate and not one to be trifled with! Olav Gundelach was, in fact, always ready to cooperate with the Union – sadly he passed away some twelve months after this meeting.

Vive le ... Pays de Galles!

Nothing enrages Welsh farmers more than French interference in the livestock export trade. The French market became an increasingly important outlet for Welsh lamb after entry to the Common Market and the highly prized Welsh product was eagerly sought by French wholesalers and their customers. However, when French workers and farmers particularly, had cause to complain, the easiest target was usually British farmers or tourists.

In one sustained campaign of direct action which brought transport to a standstill and reached a climax when French farmers broke into refrigerated British lorries and set fire to lamb carcasses, there was an angry response from Welsh farmers and a demand for some form of retaliatory action which would teach the French a lesson! The FUW complained to the European Farm Commissioner and Welsh members of the European Parliament – as it happened most were on holiday!

The French demonstrations continued – very often in front of bored French gendarmes who were clearly sympathetic and turned a blind eye to the flagrant breaches of the law occurring before them. It was then, that in desperation, I issued a Press statement urging UK shoppers to hit back by boycotting French goods, particularly French wine. The resultant publicity was gratifying but little was expected of the campaign as British shoppers were considered to be easy prey to French cuisine, brandy and wine.

To my astonishment there was an immediate reaction from some shop keepers, mostly in market towns, who indicated their support in varying degrees for the campaign. However, one notable participant telephoned to say that he would not be selling any more French wine and that he was clearing his shelves of all French products! This unexpected reaction filtered through to France and a French TV unit sped across the Channel to the shop's locality to investigate this unusual behaviour on the part of their British neighbours.

I was asked to take part in the programme and embarked on a spirited denunciation of all things French and made a passionate plea to UK shoppers to teach the French a lesson by cooperating in the boycott of French goods. It was not until I walked back to my car that I realised it was a Renault and managed to speed away without being noticed during a break in the filming!

On the Move

The Union's first headquarters was located at Gwalia House in John Street, Carmarthen, described by 'Tyddynwr' in the Cambrian News as resembling "nothing so much as the vestibule of an old fashioned French hotel." It had three rooms which were cleared and cleaned by J.B. Evans before he could move in. The offices were subsequently furnished with second-hand tables and chairs which were supplied on credit by a local store.

By May 1956 the Union had moved its headquarters to rooms above a local corner shop in Chalybeate Street, Aberystwyth. However the premises also proved inadequate and the Union moved again seven years later to larger premises at Queen's Square in the centre of the town – a building which at one time had served as a town house for a local squire and which had a pleasant view over municipal gardens near the Town Hall. The new headquarters was officially opened in 1963 by Ald. Cliff Knight, the Mayor, and served the Union's purpose for the next 36 years.

The new headquarters – Llys Amaeth – had its limitations however, notably that it had been constructed for domestic and not office use – when the time came for the Union to consider its future, the Queen's Square headquarters was clearly a non-starter The staff had become accustomed to the deranged alignment of its doors, windows and floors and assumed that most old houses full of character had their share of settlement.

My first floor room was sited above the general office and had seven heavy, metal filing cabinets ranged around the wall next to a floor-to-ceiling bookcase while I occupied a traditional, wooden desk which had six drawers usually full of papers, books etc. – quite a formidable weight in itself. There were times when publications – particularly the Union's journal, Y Tir & Welsh Farmer – were piled in any available space. Circular objects which were placed on the floor had an alarming tendency to roll across the room!

The ramifications of this 'settlement' was not taken seriously until a visiting party of local authority representatives inspected the building and wandered into my room. One tested the suspect floor by jumping up and down which so alarmed one official that he beat a hasty retreat to the door and urged the others to follow him.

Some time later I read a surveyor's report on the building which stated that the premises were never intended to be used as offices – only for domestic purposes – even then it would not meet current standards for domestic use. The structural problems, it added, had been exacerbated by additional loadings brought about by its use as offices. The report continued: "Floor structures are badly substandard even for domestic purposes." The report recommended that the timber supporting beams should be replaced by steel beams.

The floor in my room, the report added "has sagged considerably, causing distortion.... and considerable sloping of the floor surfaces... additional movement due to this weakness in the structure is likely to continue on a gradual basis and any sudden, additional loadings – for instance temporary storage of stationery or other heavy goods could lead to more serious problems." The report urged that the work should be carried out immediately.

On a more optimistic note and in an endearing turn of phrase which suggested that the building was just tired and badly in need of a rest, the report added sympathetically "Long term structural settlement of the building has occurred causing considerable distortions of the building structure. We are of the opinion that the building has reached its final position of repose and any further movement of significance is now unlikely."

It took another nine years before I, and other members of staff, moved into a new, purpose-built headquarters at Gogerddan, two miles outside the town. The building was officially opened in 1999 by the Minister of Agriculture, Nick Brown and has level floors and windows which actually open!

Recognition

From my first day with the Union I was made aware of the crucial importance of gaining formal government recognition. This would mean that the Union would have the right to be consulted on the government's agricultural policies; that it would have access to government Ministers on issues and developments in the industry and that it would he able to nominate persons for membership of government committees and bodies. Recognition would, in fact, place the FUW in the same influential position as it's rival, the National Farmers Union.

More important than anything else, as far as the FUW was concerned, was that recognition would give the Union access to the Price Review negotiations – the annual series of meetings and representations to government officials and the Minister of Agriculture on the state of the farming industry, the level of grants and subsidies, and the policy initiatives required to maintain and advance the industry's welfare. There was criticism that cereals dominated the Review – involvement in the negotiations would, it was considered, enable the Union to highlight sectors of importance to Wales e.g. the livestock sector and the hill farming areas.

The Review negotiations had a direct influence on farm incomes and every year speculation about the outcome dominated the farming Press for weeks. As the industry's representatives, the series of meetings with Ministry of Agriculture officials and the Minister, gave the NFU a valuable publicity platform which they could manipulate to their advantage. The NFU's favoured position was jealously guarded and used by its supporters in scathing criticism of the FUW's apparent inability to influence government policies. FUW membership canvassers found it particularly difficult to counter this criticism.

Prior to recognition, the FUW submitted a memorandum on the Review to the Secretary of State and the Minister of Agriculture and although its recommendations were reported in the Press it was difficult for the Union – and farmers – to know how much influence the memorandum had on the outcome of the negotiations. The first significant step on the road to recognition occurred when the Secretary of State, James Griffiths, agreed that Union representatives could meet the Minister of State, Goronwy Roberts, in March 1965 to discuss farming policies and issues. There was another significant step forward when, four months later, FUW representatives held their first meeting with senior officials at the Ministry of Agriculture. Anglesey MP, Cledwyn Hughes (later Lord Cledwyn) had played an important role in arranging the meeting and when he was appointed Minister of Agriculture there were hopes that recognition would be achieved during his term of office. At meetings with the Anglesey branch he had indicated that he would support FUW representation on government bodies. These meetings had been held in confidence but when it became clear that the Minister was not going to recognise the Union, the Anglesey branch's County Secretary, R.J. Williams warned that his previous assurances could 'leak' out to the Press.

Following the government's decision to share responsibility for Welsh agriculture between the Minister of Agriculture and the Secretary of State in the 1960s and 1970s the Union directed its Price Review recommendations to the two Ministers and held meetings on the Review with the Secretary of State and his agricultural officials. Inevitably however, the centre of attention – and influence – remained with the Minister of Agriculture and the FUW's attempts to gain recognition and access to the London Review negotiations continued unabated during this period.

The first real break-through occurred in August 1977 when the Minister of Agriculture, John Silkin stated in a Parlimentary reply to a question from Cledwyn Hughes that from 1st April 1978 the Secretary of State for Wales would exercise the same responsibilities for agriculture as his opposite number in Scotland – in effect the Secretary of State would, in reality be the Minister of Agriculture for Wales.

This announcement was warmly welcomed by FUW President Myrddin Evans but was received with mixed feelings in the NFU. Gerwyn Lewis, the NFU's Welsh Director commented: "We are not sure what the announcement will mean in practice but obviously the decision has been made for political rather than agricultural reasons."

Charles Quant, the Liverpool Daily Post's agricultural correspondent observed "For the life of me, I can't understand why the NFU Director for Wales had to give such a graceless reception to administrative decentralisation of Welsh agriculture from Whitehall to Cardiff."

The NFU's Welsh Director's reservations about agricultural devolution were not surprising in view of the strong opposition of some of his leading members. The chairman of the Union's Welsh Council, David Carey-Evans, for example, had a year earlier warned that any separate agricultural policy for Wales would create commercial barriers which would, very quickly and adversely effect farmers pockets. Such a trend, once started, would eventually grow, he added.

Dunstan Court, writing in the NFU's journal 'British Farmer & Stockbreeder' commented: "I've been amused to read of the manner in which the government is going to yield up the appearance, without conceding the reality, of devolution of some agricultural powers to the Welsh Office. For years, Fred Peart (the former Minister of Agriculture) and his predecessors, stoutly resisted any significant transfer of agricultural functions to Cardiff from Whitehall on the grounds that in virtually every field of any importance there simply had to he uniformity and that proliferation of carbon copies was in no one's interest. However the march of Welsh nationalism means that from April next the Welsh Secretary, the estimable John Morris QC will have sole authority in a number of spheres – although he may choose that jointly with his opposite number in London. It's a pretty spurious procedure though when you look closely at the list of powers to be delegated. Is it really conceivable that Mr. Morris will be able to operate the agricultural tied cottage legislation differently from England when the law is identical...? He is to be given sole responsibility for control over bees and of such significant areas as the licensing of stallions, no doubt so that he can require a minimum of three testicles before any Welsh mare can be served..."

Despite the assumption of greater agricultural powers on the part of the Secretary of State and the Welsh Office the decision to recognise the FUW when it came was totally unexpected. It was a surprise to myself and the leaders of the Union. Indeed, Myrddin Evans' first reaction when he was told was to ask whether his leg was being pulled!

The Secretary of State had, in response to another recognition request in 1975, asked for detailed, audited membership figures. These had been submitted a year later showing a total membership of 12,713 and indicating that the Union was in the majority in most of the western counties of Wales. The Union had to wait another two years before its recognition request was granted and the general view was that the Welsh Office was dragging its feet on the issue. Lord Geraint considered that the Secretary of State had little influence on the final approval and that, indeed, his early link with the Union as its Assistant Secretary proved an obstacle. As far as the Union was concerned it was generally accepted that the Secretary of State could not show undue sympathy in case he was accused of bias – however there was also the impression that the protagonists were pushing against an open door.

Various theories have been advanced for John Silkin's decision to recognise the Union. The political climate was favourable and the disposition of the leading political figures largely sympathetic. James Callaghan's Labour government was in a minority at the time and he was kept in power between 1997 and 1999 with the support of the Liberal Democrats whose agricultural spokesman was Ceredigion MP Geraint Howells, later Lord Geraint. He was one of the earliest members of the Union and had worked tirelessly to establish the Union in Mid and North Wales. He was the first chairman of the Union's influential Ceredigion branch and was later elected a Life Member after serving for 21 years on the Wool Marketing Board with the support of the Union.

Lord Geraint was proud to have played a part in the Union's recognition and often recounted an occasion early in 1978 when, after a function at the Cafe Royal in London he was invited to share a Ministerial car with John Silkin back to the House of Commons for a late vote. Lord Geraint recalled that during the general conversation the Minister raised the recognition issue and asked him: "Geraint, what do you really want?"

Lord Geraint replied: "I want full recognition, nothing more, nothing less."

To Lord Geraint's surprise, the Minister replied: "If that's what you want, that's what you'll get."

However, Lord Geraint noticed another incident at the Cafe Royal function which I firmly believe, had more bearing on the Minister's decision than anything else. During the evening Lord Geraint noticed that the Minister and NFU President Henry Plumb were having "a heated discussion."

Lord Geraint told me later – and the incident is recorded in 'A Family Affair.'*

"I never found out what it was all about but it was obvious that there had been a blazing row. Why did John Silkin raise the matter of recognition with me in the car rather than wait for a more opportune moment – after all, I met him regularly at that time."

The answer, I believe, stems from that row between Mr. Silkin and Sir Henry, later Lord Plumb of Coleshill. NFU Presidents had a close relationship with Ministers of Agriculture in the early post-war years and the arrangement benefitted both sides – indeed some complained that the NFU was an extension of the Ministry. It meant however that the Minister' s door was always open to the NFU President – that is until John Silkin appeared on the scene. Apparently the Minister found it galling to see Press reports exaggerating the NFU's influence on events. Mr. Silkin's reaction was to close the door at a time when there were a number of issues worrying farmers – the Green Pound and its impact on the government's counter inflation policy, the rating of agricultural land and extreme wintry weather which had resulted in heavy livestock losses in parts of the UK, including South Wales.

Relations between the NFU and the Minister reached an all-time low with Mr. Silkin even accusing the NFU of lacking patriotism. Sir

---

* 'A Family Affair' by Handel Jones (Y Lolfa).

Henry paid a special visit to Whitehall to ask the Minister to withdraw his accusation. The Minister refused and said at the time: "When the government says we are not going to rate agricultural land the immediate effect is – 'NFU win their campaign'. Or when Gundelach (the ECC Farm Commissioner) promises me that he will put proposals for the Milk Marketing Boards before the Council, 'NFU get MMB proposals' Quite untrue, its nothing to do with them. I think they could really concentrate much better if they were not always trying to justify their own existence."

He added pointedly "Leadership is not merely taking initiatives, though that's important, certainly not dishing out lolly to your own people, it is sometimes telling the truth – and you should always tell the truth."

Wallace Day, a respected – and restrained – agricultural columnist told me later that although he had often criticised the NFU, he was appalled by the Minister's reaction to the NFU. He described Mr. Silkin's criticism and treatment of the NFU as "the most remarkable in living memory and more objectionable than the outburst of Stanley 'Featherbed' Evans who was sacked by Prime Minister Clement Attlee from his Ministerial post after his attack on the farming industry."

Against this background of rancour and acrimony it is difficult not to come to the conclusion that the NFU President, Sir Henry, was unwittingly, the catalyst for the Union's successful request for recognition which had been formally submitted to Mr. Silkin on the 7th February 1978. The row between Sir Henry and Mr. Silkin received headline coverage in all the major newspapers and the FUW's request was granted that same month. Cause and effect? – I would say so.

Mr. Silkin's farming policies were not popular with farmers, many of whom espoused Sir Henry's reservations about the industry's plight. Like Sir Henry, their concerns centred on the Green Pound and, in direct consequence, the increase in Irish beef imports. Carmarthen FUW chairman, D.R. Jonathan had, earlier in 1978, accused the Minister of a callous indifference to the industry's problems at a Llandovery meeting. However all this was put to one side when, as the architect of the Union's recognition, he was given a hero's welcome when he attended the Union's annual meeting some weeks later as the leading guest speaker.

John Silkin was never far from controversy and later that year he stirred a hornet's nest when he suggested that the live export trade should be discontinued on animal welfare grounds despite the publication of an independent report earlier that year that came to the conclusion that a ban would not be justified on welfare or economic grounds, Mr. Silkin told a Veterinary Association conference that Britain had to look forward to a future in which there would not be a live export trade.

Recognition Repercussions

The official announcement granting recognition stated: "The Secretary of State for Wales and the Minister of Agriculture, Fisheries and Food, today announced that from April 1st, 1978, the government will formally recognise the Farmers Union of Wales as one of the organisations representing the agricultural industry in Wales. As Agriculture Minister for Wales, the Secretary of State will consult the FUW on those matters relevant to the industry in Wales as and when appropriate. He will receive and consider the views of the FUW on matters of agricultural policy for Wales and on all relevant matters which that Union wishes to bring to his attention. The Secretary of State will also, as necessary, request and will consider nominations from the FUW for appointments to bodies of which he will have a responsibility."

The statement went on to repeat the Secretary of State's request for the FUW and NFU to establish a good working relationship – he would not be able to see one Union on one day about an aspect of agricultural policy and see the other Union the next day on the same issue.

The statement was received with enthusiastic acclaim by the FUW. FUW President Myrddin Evans had been telephoned with the news the previous night and regarded it as the most memorable moment of his 18 years in office. He recalls that when a senior official at the Welsh Office phoned him with the news he thought for a moment that a colleague was pulling his leg.

Myrddin Evans considered it was a historic moment for the Union and Welsh agriculture. Referring to the Price Review negotiations he added "We have had discussions with the government for a number of years but we have always had to go in by the back door – now we can go in by the front door with our heads held high."

He also recalls that I chastised him for wearing a polo neck shirt the next day instead of the usual collar and tie after arrangements had been made for representatives of the agricultural Press to interview him on the news. A member of staff was hurriedly despatched to the nearest shop to buy a shirt and tie for the photo call. That day's meeting of Union leaders was stunned when they heard the news and Vice President Megan Davies of Montgomery promptly burst into tears.

The news was toasted with the remainder of some champagne that had been presented to the Union by its insurers, National Employers Mutual, to mark the Union's 21st birthday.

The announcement was greeted by an avalanche of congratulatory messages from a variety of supporters and sympathisers. Equally impressive was the tidal wave of congratulatory leading articles in the Press, most of which considered that recognition was the prelude to the unification of the two farming Unions in Wales.

> *Now that the FUW has won government recognition, it would seem logical that the NFU's Welsh membership and the FUW should join forces in a single Welsh Union which would fulfil in Wales the role that the Scottish and Ulster farmers Unions fulfil in their own corners of the UK.* (Big Farm Weekly)

> *Perhaps the best solution which could accommodate the aspirations of both Unions would be the creation of a Welsh Farmers Union joined to the NFU in exactly the same way as the Scottish and Ulster farmers Unions which maintain their separate identity while being joined in a federal structure with the NFU in England and Wales. This would allow full freedom for Welsh farmers in negotiations but also give them access to the power and facilities of a much larger organisation which deploys resources in London and Brussels.* (Western Mail)

> *There is no reason why a separate Welsh Farmers Union on the lines of the Scottish NFU should not be set up.* (Farmers Guardian)

> *The announcement by the Ministry of Agriculture that it has recognised the FUW is in keeping with the fashion for devolution and, in my view, no bad thing. Welsh farming has a character and special requirements all of its own because of its small farms, wet climate and rural setting.* (CW Scott, Daily Telegraph)

A student of the FUW's early history is left with the clear impression that if the policy makers at Agriculture House (NFU headquarters) had been more sensitive to the needs of Welsh farmers, the FUW would not be the force it is today. (The Times)

For 22 years, farming in Wales has been represented and served by two organisations; the NFU and the rival FUW. The NFU 'split' which brought about the second force came about amid argument and acrimony. Since then the FUW, as a radical, campaigning break-away movement has shown little respect for the 'Big Brothers' of Knightsbridge, Westminster and Whitehall. Its appeal to the many who find the Establishment out of touch with the needs of farming's little men has been real and often under-rated. (Farmers Weekly)

The FUW is to be congratulated on the successful outcome of nearly a quarter of a century's campaign. (Liverpool Daily Post).

The officials of the FUW are naturally jubilant at being brought in from the cold to official status and it is to be hoped that for the benefit of both Unions, and the members they represent, and for the ultimate good of the industry in Wales – that the road is now clear for cooperation between them to take a turn for the better. (Cambrian News)

In an otherwise critical leading article, even the NFU's journal, the British Farmer and Stockbreeder paid grudging congratulations to its Welsh rival while NFU Welsh Council chairman, Tom Rosser expressed his good wishes,

The new spirit of good will did not last long. As the FUW prepared for its first experience at the Price Review negotiating table, the NFU dropped a bombshell – NFU President Henry Plumb announced that the NFU delegates at the negotiations would not accept the presence of the FUW team!

Sir Henry stated that the two Unions would have separate Price Review talks until "a settlement was reached in Wales". Sir Henry's threat became a reality when, at the first scheduled London meeting with government officials to discuss the EEC's sheepmeat regime, the NFU refused to attend and insisted on separate meetings.

I immediately issued a statement on behalf of the Union emphasising that if the NFU continued with this policy it would be a tragedy. The statement continued: "The FUW has acted in good faith and considered it had moved into a new phase of trust and cooperation with the NFU. This action will sour relationships between the two sides and jeopardise the new spirit of cooperation in Wales. It will be quite easy to revert to the open warfare of the past but we don't think it will be in the best interests of Welsh farmers."

Myrddin Evans described the NFU's action as 'childish' and said that the NFU seemed to be more interested in protecting the Union's status rather than the farming industry.

In a statement issued in April of that year, Sir Henry poured cold water on any hope of reconciliation "The NFU is in Wales to stay and we shall continue to strengthen our position as the voice of the agricultural industry." Sir Henry added that he had attended a meeting of the Welsh Council of the NFU and he added "...the message is clear... what we cannot do, and this is the opinion also of the NFU of Scotland and of Ulster Farmers Union, is to commit ourselves to ending overnight the understanding and arrangement which has been built up over 30 years."

The Secretary of State, John Morris, voiced his disapproval of the NFU's decision and added: "I hope on mature reflection it will be recognised what a nonsensical situation this is. We have had to duplicate meetings dealing with precisely the same items. It is a matter of regret that the NFU has taken this attitude – it is a waste of time."

Montague Keen, in an Editorial in the NFU's British Farmer and Stockbreeder, went into more detail and stated that both the Scottish and Ulster Unions had objected to the presence of FUW delegates at the Price Review and Ministerial meetings. He wrote: "In fact, it was apparent that the Scottish and Ulster Unions had objected just as strongly to any question of the four-party discussions in London without having established quite clearly the basis on which any FUW participation might take place."

Montague Keen added that the Scottish and Ulster Unions had objected on the grounds that they had, over the years, contributed towards the cost of representation and common services. He added "Would it be reasonable to expect overnight participation by a body (FUW) which they considered unrepresentative, which had made no approaches on the principle let alone the mechanics of cost sharing, or on the formulation of a common policy? It would be quite unrealistic to think that the FUW, after 22 years in the wilderness, could expect to slip quietly into this harmonious arrangement in which the cause of Ulster, Scottish and Welsh farmers are vigorously advanced, yet within an agreed framework of a common approach."

As FUW President Myrddin Evans had indicated quite clearly on numerous occasions that the Union would finance any representation at Whitehall or Brussels, the reference to representational costs mystified the FUW.

The inclusion of the Scottish NFU in Sir Henry's reasons for refusing access to the FUW was ironic as only a year earlier the Scottish NFU had stated proudly that it was "jealous of its independence and insistent that it should play an effective role in London and Europe" – but it would not neglect its domestic affairs. This was exactly the role that the FUW sought and was, apparently, being denied by the Scottish NFU.

When I issued a statement to the Scottish Press emphasising the apparent inconsistency of the Scottish NFU's position, the Scottish Farmer reported that I had launched "a blistering attack" on the Scottish NFU. Michael Burnett, the Scottish NFU President, said he was baffled by my comments and laid the blame for the row on the government and added: "Our present understanding is that government recognition of the FUW referred only to contact between the FUW and Secretary of State."

This explanation beggars belief in view of the fact that the Ministry of Agriculture's decentralisation statement in July a year earlier had referred specifically to the fact that the Secretary of State for Wales would exercise exactly the same responsibilities as the Secretary of State for Scotland – thus recognition for the FUW would confer on it the same status as the Scottish NFU. There was never any suggestion that decentralisation in Scotland would confine the Scottish NFU's representations to the Scottish Office and the Secretary of State for Scotland!

The Scottish NFU's journal, the Scottish Farming Leader stated that it was quite wrong for the government to introduce "a completely new organisation to join the UK farmers Union team" at the Price Review negotiations. It admitted that the Union was concerned about costs and finance... "we carry our full share of the costs involved in playing our part in agricultural policy negotiations at UK national and international level and these are issues that have to be resolved..."

The Farmers Guardian described the NFU's decision to boycott the meetings with the FUW as "the most extraordinary mistake" and added: "The unpleasantness generated by a refusal to sit alongside the FUW at a meeting with the Ministry is likely to sour relations between the two organisations just at the very moment when there should have been hope of progress towards cooperation which would have benefitted the whole of Welsh agriculture."

A suggestion that both sides should sink their differences and reach an understanding was mooted and pursued in talks which ultimately ended in failure – and further acrimony – a year later. FUW leaders considered the talks were a smokescreen and that there was little hope of achieving a fair settlement.

Despite Sir Henry Plumb's role in denying access to the Price Review negotiations there was a strong impression amongst FUW leaders that he was reflecting the views of a diehard element inside the Welsh NFU. FUW President Myrddin Evans met him on several occasions at this time and retained his respect and admiration of the NFU leader. Myrddin Evans commented: "Sir Henry was always willing to discuss proposals constructively – he did not attempt to impose his views on one. He was always fair and we respected each other."

The FUW was never allowed to attend Price Review meetings with the NFU but by this time the negotiations had been relegated in importance following membership of the EEC and had become subsidiary to representations in Brussels. The Conservatives regained power in 1979 and although the new Minister of Agriculture, Peter Walker, urged the two Unions to sink their differences, the departure of John Silkin and John Morris reduced the urgency of the issue. It made little difference to the FUW – it was now a recognised body and its first Price Review team, comprising Glyngwyn Roberts (on beef), H.R.M. Hughes (milk), Ieuan Davies (sheep), Michael Dolan (pigs), Irwyn Phillips (cereals/potatoes), Tom Jones (Montgomery) and R.Ap Simon Jones (hill farming and marginal land) led by Union President Myrddin Evans, and accompanied by headquarters commodities staff and General Secretary Evan Lewis, attended their first Review meeting at the Ministry of Agriculture in December 1978.

As far as Brussels was concerned there was never any problem – the Union had received an assurance from Peter Parkhouse of the EEC's Agricultural Division and Gwyn Morgan of the EEC's Welsh Office that it would have direct access to the decision-making process in Brussels. This proved to be the case. As far as COPA (the European Farmers Association) was concerned the Union never regretted its non-membership and noted COPA's own admission that as much work was done by individual countries in consultation with their own Ministers as by COPA itself.

Council Cameos

FUW Council meetings were sometimes dull and sometimes dramatic occasions covering a wide variety of topics ranging from complex UK and European rural policies to domestic administrative issues. Sometimes serious discussions were enlightened by flashes of unintentional humour but that was certainly not the case in 1957 when the Council erupted in the furore which followed the London meeting between J.B. Evans and D.J. Davies with NFU President Sir James Turner and four members of the NFU Welsh Committee to discuss unity.

In a prepared statement D.J. Davies told Sir James: "I believe that the division of Welsh agriculture can only be resolved by a bold statesmanlike act by you, an act which by its breadth of vision and generosity, will capture the imagination of all Welshmen, an act which will wipe the slate clean so that a fresh start may be made in Welsh agricultural organisation." He went on to suggest that Sir James should call the two sides together with the objective of establishing a new organisation in Wales with the same powers and status as the NFU's of England, Scotland and Ulster.

D.J.'s eloquent plea fell on deaf ears and some days later he and J.B. Evans were pilloried at an emergency meeting of the Council at which the doors were closed and delegates were not allowed to leave or enter the stormy meeting. A leading member of the Union, Captain H.R.H Vaughan resigned because the resignations of D.J. Davies and J.B. Evans had not been accepted. Later, on 3rd January 1958 another meeting of the Council was told that the decision had been reversed at a committee meeting that morning and the Council duly accepted the two resignations along with that of another founder member of the Union, Llew Bebb of Goginan. D.J. Davies left the meeting amid uproar after making a plea on behalf of J.B. Evans. J.B. Evans' resignation was subsequently deferred while Llew Bebb and Captain Vaughan were persuaded to return to the fold. D.J. Davies however spent some years out of the Union before he was reinstated as a life member.

Price Review Reproach

The 1960s and the 1970s were periods when unpopular Price Reviews and general pressures on farm incomes resulted in stormy Council meetings. Sometimes the government's senior civil servant – the Welsh Secretary at the Welsh Office Agriculture Department – was invited to attend Council meetings to defend and justify government policies. One notable Welsh Secretary who braved this ordeal was Hywel Evans, former head of the agricultural advisory service in Wales. He did not look like a desk-bound civil servant – a tall, commanding figure dressed usually in tweeds he was a bluff, no-nonsense character who had a reputation for colourful language.

One of the meetings he attended occurred after one of the most controversial Price Reviews of that era when government ministers infuriated the farming community by claiming that the result of the Review represented a cost-plus outcome for the industry. Despite a barrage of criticism Hywel Evans stuck manfully to his task of defending the government. But – clearly impatient after one excessive criticism he shook his head disbelievingly and muttered "Christ, Almighty". At this point – and in the silence that followed – a figure rose from the back of the meeting and walked slowly and deliberately up to the speakers rostrum and stood in front of the Welsh Secretary. Delegates held their breath as the elderly farmer wagged his finger in the face of the incredulous Welsh Secretary and declared loudly "Don't you dare take the name of the Lord God in vain." Amidst the silence he turned on his heel and walked slowly back to his seat. The President, Myrddin Evans, moved the discussion on while the bewildered Welsh Secretary mopped his brow and continued his stubborn defence of the government's policy.

The King is dead ...

Staffing issues sometimes posed more problems at Council meetings than agricultural policies especially when delegates challenged the decisions of the Finance Committee which comprised the leaders of the Union. Staff unrest reared its ugly head in 1993 when a promising staff evaluation and salary scheme was introduced but ran into difficulties only two years later. I was elected chairman of the Union's Salaried Staff Association at this time and often took advice from a helpful ACAS (industrial arbitration) source. The staff unrest was compounded by other changes in the Union's administrative structure. These problems were, inevitably, raised at a stormy Council meeting where FUW President Bob Parry, who was also chairman of the Union's Finance Committee, defended administrative changes against some critics, one of whom accused him of double standards.

At an early stage in the discussion the staff were asked to leave the meeting and I accompanied them to an ante room where we settled down to wait for the outcome of the debate. After a few minutes I was asked to return to the meeting to take notes and write the minutes. The acrimonious debate continued until the President, who was chairman, lost patience and announced to shocked delegates that he'd had enough and was resigning! He gathered up his papers and marched out of the meeting, through the ante room where he told the staff he had resigned as he sped through to his car in the car park.

I was appalled at the public relations consequences of this incident and raced after the President. As I hurried through the ante room I took in the shocked faces of the staff, one of whom broke the silence with the words " The King is dead, long live the King..." I reached the President's car and appealed to him to return to the meeting. I was soon joined by an Anglesey delegate who joined in my appeals. To our relief he relented and did so – the meeting was suspended after passing a vote of confidence in the President and Chief Executive, Roland Williams. The Council meeting was reconvened three weeks later and the recommendations which had been the subject of controversy at the previous meeting were confirmed and approved.

Foreign Service!

Scandinavian tour – Agricultural journalists from the home countries were invited to tour Scandinavian farms and meet farming representatives in 1993. Here we are on a Norwegian farm with journalists from Ireland and Ulster.

A Brussels line-up when North Wales MEP Beata Brookes – a formidable advocate – succeeded in 'extracting' Farm Minister Gundelach (centre) from 'an important meeting' to discuss farm policies with FUW President Myrddin Evans.

The FUW has always enjoyed a good relationship with its Irish neighbours – here the Irish agricultural attache enjoys a joke with Bob Parry and Gwilym Thomas.

Brussels Riot – face to face with the law in Brussels in the form of a bad tempered riot policeman!

FUW delegates led by Myrddin Evans and H.R.M. Hughes managed to avoid the worst of a clash between riot police and steel workers when they met EU officials at EU headquarters in Brussels. A number of police were injured when they charged the demonstrators.

H.R.M. Hughes greets a group of Spanish farmers at Aberystwyth.

On home soil ...The FUW has forged a close link with the Welsh Assembly and its Rural Affairs Minister. This picture was taken at Vice President Glyn Roberts' farm near Betws-y-coed. Left to right: Bob Parry (President), Gwilym Thomas, Glyn Roberts, Gwynedd Watkin and Carwyn Jones AM.

A War of Words

Favourable publicity was regarded as the essential ingredient required to fuel the development of the FUW in its early days. Leading founder members of the Union, notably D.J. Davies, complained that the Union would never attract new members unless the farming community was aware of its policies and representations on behalf of Welsh farmers. He complained that some national and weekly newspapers failed to acknowledge the existence of the Union – one prominent West Wales newspaper which served a large farming area, refused to publish reports about the Union despite its controversial establishment. D.T. Lewis, the Union's second President, complained bitterly that he had been ignored by the Editor of Welsh Farm News when he addressed a young farmers meeting in Brecon.

Several attempts were made by the Union to acquire professional assistance but it was not until nine years after its establishment that the post of Public Relations Officer was advertised and I was appointed. I soon discovered that I was expected to fulfil a variety of other duties, including for a period, as Secretary of the Union's main policy committee. Nevertheless, from the outset I decided to give priority to Press Releases circulated to daily newspapers, radio and television, the farming Press and weekly newspapers.

My second priority was to keep members informed of the work which the Union was doing on their behalf and this meant the production and distribution of a regular journal. At that time the Union produced an occasional tabloid-sized journal ' Y Tir' which was printed locally.

The Union's precarious financial position made it vitally important to attract advertising to cover the costs of publication and postal distribution. When I joined the Union the responsibility for advertising procuration was that of a Cardiff advertising agency. When I told the Union's Finance Committee that I was not impressed by their services I was told to take over the work myself and had no alternative but to do so! For the next 38 years when I produced and edited 'Y Tir' its financial fortunes fluctuated. Its title was amended to include Welsh Farmer in order to encourage more advertising – its life blood. After a period in magazine form it settled down as a tabloid monthly with a circulation in the region of 15,000. I resisted the temptation to produce a publication of pure propaganda in the belief that unless it covered issues of general and immediate interest it would not be read. The inclusion of articles by invited columnists introduced external expertise and some variety, and proved popular. One observer commented that I succeeded in transforming 'Y Tir' into one of the best farming publications in the U.K. The best proof of approval was that a number of NFU members requested copies of the journal.

The expansion of radio and television in the late 1980s and 90s – and the succession of adverse, but newsworthy developments which hit the farming industry, placed intolerable burdens on the Union's publicity services and the Union's journal may well have suffered had I not had the good fortune to contact a representative from a West Country publisher, Community Media Limited (CML). The subsequent negotiations, which resulted in an agreement which transferred the printing and advertising work to CML and left the Union responsible for the editorial content, proved a very satisfactory outcome which, from 1992, established an effective working relationship which lasted for many years. A similar arrangement was later established with South West Wales Publications of Swansea.

As far as publicity was concerned I soon discovered, to my astonishment, that some newspapers did not treat FUW Press Releases on their news value and wilted under external pressures from our competitors. This fact was noted in an editorial in the Cambrian News a few months after the establishment of the Union. The editorial by Tyddynwr noted: "What I have found hard to accept is the attitude of certain elements of the NFU. If they had their way all discussion on the new movement would be silenced; the Press gagged and the discontent of the farmers denied a hearing...no one has the right to deny the FUW a fair hearing."

But try they did – one agricultural correspondent on a daily newspaper told me that after publishing a report from the FUW he was criticised by a prominent NFU member for giving publicity to the FUW and told that unless it stopped the Editor would receive a delegation and be told in no uncertain terms that the paper would have to revise its farming coverage. Fortunately, the reporter resisted this pressure and indeed, in some cases it was counter productive.

Some Editors responded positively to complaints of bias. One example involved a North Wales election which was a straight contest between an FUW and NFU member. The two candidates farmed within a short distance of each other and, in fact, were on the best of terms. However the election was, as usual, regarded as a trial of strength of the two Unions. A local newspaper included an eve-of-election article on the contest and the reporter – whose impartiality had been suspect in the past – concentrated solely on the NFU candidate and totally ignored his FUW opponent. The Editor was horrified when this injustice was pointed out to him and spent no time in transferring the reporter to other duties.

Elections to the Marketing Boards were – as indicated earlier – regarded as a test of strength of the two Unions. Like all other members of staff I was involved in numerous election campaigns, some successful, some unsuccessful. The most significant break-through occurred in 1962 when FUW candidate Evan R Thomas of Llanybri won the coveted South Wales region Milk Marketing Board seat by defeating NFU member Bill Hinds whose family had held the seat since the formation of the Board. Bill Hinds regained the seat three years later but the NFU's reputation of invincibility had been shattered in one of the most important milk-producing areas in the UK.

Roger Evans of Carmarthen emulated Evan R. Thomas' success years later and became a highly respected producer representative who transcended Union affiliation. Similarly, Geraint Howells, later Lord Geraint, was elected the Welsh member of the Wool Marketing Board in 1965 and served with distinction for 21 years.

One huge thorn in the Union's side until it ceased publication in 1969 was Welsh Farm News, a weekly agricultural newspaper published by Caxton Press of Oswestry which was financially supported by the NFU and editorialised accordingly. The Union's main critic was its columnist 'Ben Brock', a pseudonym which hid the identity of a well-known West Wales journalist. Ben Brock's colourful criticism of the Union can only be described as vituperative and he produced blistering broadsides on the FUW with unfailing regularity.

In one response to a letter of complaint from a North Wales FUW member, Ben Brock retorted: "It has always been abundantly clear that I am not of his camp, colour or creed for the simple reason that I believe his organisation does nothing but the greatest harm to our industry which the NFU serves to the best of its ability" and referring to the correspondent's plea for one Union for Wales, he replied: "How disillusioned and tragic he and his fellows must feel that after ten years of rancour, mud slinging and beating the air, they still don't have one and, what is more, are further away than ever from having one."

Ben Brock's main weapon was ridicule and his New Year almanac column always included a humorous and offensive swipe at the FUW. I came to the conclusion that the only way to combat this was to utilise the same weapon and became engaged in a war of words on the virtues of the FUW in the correspondence column. In his column in March 1965, Ben Brock finally commented: "It appears that G. Lloyd Thomas has joined the ranks of our contributors and we must bid him welcome... this boy is good, in fact, he's very good, we must keep him in the act – welcome fellow clown."

In my response – at a time when Bill and Ben were at the height of their popularity on children's television and there was increasing criticism about the boredom and strictly regulated annual meetings of the NFU, I responded that despite my determination to ignore his column his lavish praise had been too much for me and continued "He says I'm good… in fact very good. Good heavens lad, steady on. He wants to keep me in the act. Well, alright, if he must. But I'm prepared to share the limelight. What about a double act. Now, let me see... Brock and Thomas, no... I know, I've got it, Gwilym, in English its William – cut it to Bill and we have it... Bill and Ben... agents please note we're on TV most days. Ben is the one who keeps falling on his face. First public appearance – next year's annual meeting of the NFU... anything to brighten it up a bit."

So it went on with both of us attempting to score points off the other. Ben Brock could be a savage critic – when Emlyn Thomas, who had distinguished himself opposing the Rural Development Board at an Aberystwyth public inquiry, left the Union for a post with the Liberal party in Wales in 1968 he commented: "It must be a matter of very small national concern and of interest to only a limited number of people that Emlyn Thomas has teamed up with the Liberal party. No doubt, people will remain Liberal in spite of this... Mr. Thomas has always been reckoned to be a very able man; certainly I have never questioned this. But I would now pose the question that if he sees the prospects for the Liberals in Wales as better than those of the outfit with whom he has more recently been associated, then isn't it about time to join in prayer?"

Thankfully the paper ceased production in 1969 and some years later when I met Ben Brock at an agricultural show I found him to be a pleasant character with a good, if idiosyncratic sense of humour.

A fast and efficient service for the Press continued to be my top priority despite numerous distractions. My main link was through the distribution of Press Releases although for most of my period with the Union I also wrote and supplied articles for other publications – fortnightly for the Western Mail's farming supplement and occasionally for Livestock Farmer and similar publications.

Frequent Press Releases imposed a heavy burden on the Union's secretarial staff – at one time they had to be typed, duplicated and posted the same day to around a hundred recipients with varying degrees of urgency – TV and radio and daily newspapers, agricultural journals and weekly newspapers, etc. Press statements were also sent out in the Welsh language to those who wanted them. The demands made upon an incredibly willing and efficient administrative staff can be guessed at when around 300 different Press Releases were issued in a busy year. More modern methods of communication, fax machines and the internet, have considerably eased the administrative burden.

A similar revolution in TV and radio facilities has resulted in less demands being made on participants on news and current affairs programmes. For far too long the BBC had only two television studios in Wales, in Cardiff and Bangor, while the only ITV studio was in Cardiff. For the 18 years that Myrddin Evans was President, I invariably accompanied him to Cardiff when he appeared on news programmes. It necessitated a three hour drive to Cardiff along the A48 – the M4 was built later – sometimes for a 2½ minute interview! Early evening news programmes were usually followed by news magazine or current affairs programmes like 'Heno' and such journeys were considered very worthwhile if one was 'lucky' enough to appear on more than one programme, or on both channels. Very occasionally disaster struck when the news item involving the FUW representative was cancelled for some reason! Fortunately for me, Myrddin Evans' reign as President was followed by that of two Anglesey Presidents, H.R.M. Hughes and Bob Parry, who only had a short journey to the BBC studios at Bangor.

My Cardiff journeys strengthened my view that there should be better radio and TV facilities in Mid Wales and several approaches were made to politicians and local authorities to support the Union's campaign. The BBC established a studio at Carmarthen which eased the situation but nevertheless a large area of Mid Wales remained uncovered. I approached the University at Aberystwyth in 1995 to ascertain whether the college's proposed media centre could house a small studio or inject point and received an encouraging response from the college and the BBC. Aberystwyth now has this facility and the Union can at least take some credit for this development.

Requests to supply a spokesman for a radio or TV programme sometimes posed particular problems. Ideally the interviewee would have personal experience and knowledge of the issue to be discussed while, at the same time, be able to reflect the angry mood of the farming community – whilst retaining his self control. Any number of factors can affect the interviewee's response to questions and there is never any guarantee that the interviewee will do himself – or the Union – justice. The interviewer's nightmare is a monosyllabic interviewee who answers every question with a hesitant 'Yes' or 'No'. Fortunately, such experiences were rare at that time because the Union had considerable talent within its ranks in that respect.

Appearances on TV or radio interviews were so important to the Union's image that occasionally experienced broadcasters and journalists were called in to advise Union spokesmen on TV and radio techniques and subject them to hostile interviews.

One took a calculated risk when recommending participants for some programmes. During the stormy 1970s when economic pressures resulted in angry meetings and militant demonstrations, TV and radio producers wanted farmer participants for programmes reflecting the industry's problems and militancy. In order to have a spirited and lively studio discussion, participants had to have strong views and be able to express their anger in a coherent manner. On most occasions the result was compelling viewing but on one notable, never-to-be-forgotten occasion it went disastrously wrong.

On this occasion the BBC asked me to 'supply' a group of farmers for a studio discussion on the problems in the industry. The programme was live and was hosted by Vincent Kane. As the programme opened the farmers sat in serried ranks listening carefully as Vincent Kane introduced the programme and outlined the background to the stormy events in the industry. At this stage one member of the audience disagreed with some aspect of the opening remarks. He stood up and walked from his seat to remonstrate with Vincent Kane. Despite repeated appeals he refused to return to his seat and the programme continued in disarray until the end. I watched in horror, not only because he was one of the group I had recommended but also because an opportunity to explain the industry's problems to the general public had been wasted.

The degree of support by Union members to organised protests was also always a worry. When I organised protest marches through Cardiff in 1974, and again in 1996, I was never sure how much support would be forthcoming despite the strong feelings in the industry. Most of the Union's members had small, family farms who could not afford to lose a day's work. But the response on both occasions was heartening – in 1974 the protestors cavalcade – in a variety of modes of transport, was over a mile long. Both protest marches ended up at the Welsh Office in Cathays Park where a large crowd was addressed by prominent politicians.

Livestock Exports

The decision by the cross-channel ferry companies to ban live animal exports in 1994 plunged the farming industry in Wales into a crisis which the Union was determined to counter, one way or another. The ban was introduced despite the fact that government inquiries had exonerated UK livestock transporters of causing unnecessary suffering. Although the Union supported efforts to improve livestock transport conditions it believed that the worst excesses occurred on the Continent and that Continental companies should emulate the high standards set in the UK. One obvious anomaly was the fact that veal crates had been banned in the UK in 1991 on welfare grounds but they were still in use on the Continent.

The implementation of the cross-channel ban had a disastrous impact on lamb and calf prices at Welsh markets. The High Court was told that since the ban, calf exports, usually worth £95 million a year, had fallen by 50 per cent and lamb exports by some 70 per cent from £80 million. The Union recognised that the issue was one of vital importance to farmers in Wales and set about finding its own solutions to the impasse.

I was involved in a number of initiatives to find alternative means of continuing the trade – contacts were made with new companies interested in shipping livestock to the Continent and strong representations were made to the government – including evidence on the issue to the Welsh Affairs Select Committee. The Union pressed the EEC Farm Commissioner to introduce improved welfare standards on the Continent. Financial support was given to Gloucester livestock exporter Peter Gilder who subsequently won his High Court action against Dover Harbour Board which had discontinued live exports.

Before this victory I was contacted by an Irish livestock exporter who agreed to fly calves from a suitable airport in Wales to the Continent. Swansea airport appeared to have all the required facilities and with the approval of the airport operators, arrangements were made for the Irish company to fly in a suitable aircraft for this purpose.

Subsequently a Russian Antonov plane landed at Swansea on 10th January 1995 which was capable of carrying up to 150 calves on ten flights a day. The aircraft had a pressurised cargo area and arrangements were made for every flight to be accompanied by a veterinary surgeon to ensure the highest possible welfare standards. As it turned out the development was accompanied by lurid headlines in the national and local newspapers. I organised a Press conference to explain the welfare advantages but the local newspaper's editorial thundered 'Airport must not deal in blood money' and went on to describe the trade as 'disgusting'.

The Press headlines ensured that the first consignment of calves from Carmarthen market had a hostile reception from an increasing number of protestors. Despite the illegality of the protestors' actions their blockade attracted the support of local dignitaries and political leaders – and the apparent indifference of the police. One local MP demanded strong government action to stop the trade while another commented: "The manager at the airport should know that this is a cruel, unnecessary trade." Another spoke out in support of the protestors' actions and volunteered to represent them.

Genuine attempts were made to meet the objections of the protestors – however, despite assurances that the calves would travel in the best possible conditions and that the total journey time would be reduced to four hours the opposition increased and appeals for the police to protect the first lorry load of calves and ensure access on the public highway to the airport fell on deaf ears. By this time protestors had surrounded the stationary lorry and some of the protestors had chained themselves to the back axle.

During all this time the Union had the greatest difficulty in dissuading groups of farmers from travelling to the airport to 'free' the lorry. The Union was inundated with these offers of assistance but was faced with the nightmare scenario of an escalating war of words which would inevitably result in physical confrontation – a scenario that the local police and dignitaries seemed prepared to ignore. The stalemate continued and the tension was only relieved when the Union decided, in conjunction with the Irish exporters who owned the plane, that it should fly back to Ireland. The lorry was freed and returned with its load of calves to Carmarthen. In view of the failure to protect the lorry it was decided not to proceed with alternative proposals to ship livestock out of Swansea or Port Talbot docks; a proposal to use Withybush airport at Haverfordwest was also deemed to be impractical.

The High Court's subsequent ruling on bans on live exports at Coventry Airport, Dover harbour and Plymouth docks had lessons for what happened at Swansea. The port authorities, it ruled, were 'guilty of surrendering to illegal mob rule.' The three authorities involved were criticised for failing to recognise the dangers of 'surrendering to the dictates of unlawful pressure groups' said Lord Justice Brown.

He continued: "The implications of such surrender of the rule of law can hardly be exaggerated – if ever there were cases demanding the court's intervention in support of the rule of law, these are they." Lord Justice Brown added that the effect of the bans on beef and lamb prices was, for many, devastating. The farming community, he added, was facing a crisis with many likely to go out of business. Following the court's findings, Peter Gilder's first consignment of livestock left Dover with the protection of a strong police escort without major incident.

A Welsh initiative, Farmers Ferry, which was supported by the FUW, established a new cross-channel service between Dover and Dunkerque. The new operators established stringent welfare conditions and transported thousands of sheep which generated an income of around £1 million for UK producers every week.

The Union's involvement and initiatives during this period reaped significant support and expressions of gratitude from farmers across the UK. FUW President Bob Parry commented: "There is no doubt that the Union was the most active defender of farming's interests; we dictated events rather than reacted to them as others did. We were instrumental in providing new outlets for our stock – whatever happens in the future, we shall continue to defend our right to pursue a legitimate trade."

The Farmers Weekly commented at the time: “At the end of a year during which the FUW enjoyed unprecedented national publicity, President Bob Parry and his executive team are certain to be acclaimed for their uncompromising high profile defence of live exports... in recent years Welsh farmers have been hard put to identify many fundamental differences between the policies of the two Unions, or how they lobbied in support of them – since last August the FUW has pulled out all the stops to prove that it speaks for more livestock farmers than its rival. It’s leaders can set out to their annual meeting safe in the knowledge that they have won another propaganda war, hands down.”

And the Western Mail of January 1995 commented: “Seasoned agricultural journalist Robert Davies, who has observed the workings of the FUW at close hand said: ‘On the live export issue the FUW has led the field and I find it extraordinary that a Welsh Union should get the enormous amount of publicity it has had. There is little doubt that many farmers see the FUW as speaking for the industry on live exports, both in its current attempt at airlifting calves from Swansea and in supporting alternative cross-channel services in the wake of the ban on live exports to the Continent by the major ferry companies’...”

Despite the FUW’s efforts to counter the livestock export ban; despite the plaudits it earned and the criticism that was directed at the NFU on the issue, an effort was made by the FUW to present a united Union front on an issue which it considered to be of vital importance. Although a meeting of Union leaders at Bala was described as ‘cordial and constructive’ little came of it and soon afterwards the NFU accused the FUW of employing ‘megaphone tactics’.

Bob Parry commented later, “Our efforts showed Welsh farmers that we are a Union which fights on their behalf. The export of livestock is a legitimate trade which is of enormous importance to Welsh producers”. He added that the Union fully supported new rules and regulations which which were introduced in July 1995 for the welfare of livestock in transit and he called for more stringent welfare and transport rules on the Continent. The Union also campaigned for a Common Market ban on veal crates which had been discontinued in UK farming.

The overwhelming majority of farmers are in favour of high animal welfare conditions and the FUW played its part in ensuring that other EU countries introduced new methods of production. When new UK livestock transport regulations came into force, FUW President Bob Parry commented: "The new regulations represent a significant advance in livestock transport standards – I trust that animal welfare organisations will recognise the efforts we have made to meet their reservations."

'Mama-Miah!'

A number of Ministers who had responsibilities for farming were from urban backgrounds and constituencies and some were honest enough to confess that, initially at least, they knew very little about farming. One of these was George Thomas, Secretary of State for Wales and later Speaker of the House of Commons who overcame a strong urban image to become very popular with FUW delegates. Glyngwyn Roberts thought highly of his attentiveness and patience while General Secretary Evan Lewis recalled that he encouraged farmers to detail their problems and views. He was a good listener and courted popularity with his informal approach.

Welsh agriculture went through a difficult time during his period at the Welsh Office but his popularity survived these crises.

George Thomas reacted strongly to the FUW's decision to withdraw its annual meeting invitation to Lord Cledwyn after the 1970 Price Review and cancelled arrangements for a London lunch with the FUW President, Myrddin Evans. He was also displeased at the Union's strong opposition to the Mid Wales Rural Development Board. Speaking at the Union's annual meeting in 1969 he said he saw the Board as an instrument for shaping the long-term prosperity of Mid Wales. He was told by President Myrddin Evans that the Union would stick to its guns and continue to oppose the Board. The government subsequently announced that it would go ahead with the Board in a modified form but the Union, with the strong support of the Country Landowners Association, fought a delaying action which included a Parliamentary petition and the Conservatives returned to power and fulfilled their promise to scrap the Board.

One of George Thomas' endearing traits was his reported attachment to his mother who, it was said, sometimes accompanied him on his official duties. This was the subject of a lot of good humoured banter in the Union's staff-room before his attendance as the guest speaker at the 1969 Annual meeting. Suggested arrangements for his mother were all dismissed as fantasy as her presence at a farmers gathering was considered to be highly unlikely. Nevertheless when the Minister's limousine came to a halt outside the Aberystwyth sea-front hotel where the official lunch was being held, I welcomed him as he stepped out and was about to close the car door when he mentioned that his mother was in the car. I looked into the darkened interior and to my surprise was greeted by a smiling, white-haired elderly lady with what appeared to be a shawl around her shoulders. I transferred both to the care of General Secretary Evan Lewis who was told that she would be perfectly happy to have a cup of tea in the hotel lounge during the official lunch and go shopping in the afternoon!

I met George Thomas at a Machynlleth function some time later and he continued his criticism of the Union's failure to see any advantages in the Rural Development Board. He shook his head and had a twinkle in his eye when he said that, like miners, farmers were sometimes headstrong creatures and that he had sometimes wished that he'd had a few sheepdogs to keep them on the right track!

An Offer One Couldn't Refuse?

The 1978 Royal Welsh Show brought the usual problems and challenges for those, like myself, who were involved in preparing a trade stand for the annual invasion of members and dignitaries. The stand – on the main ring and therefore on one of the main thoroughfares – was little used between shows and usually needed a belated spring clean and sometimes, redecoration. The usual team of head office staff descended on the stand and always succeeded in creating a favourable impression for visitors. As far as I was concerned it was a question of preparing promotional material for the publicity stands which were sited in the lounge/reception area and trying to prepare for the unexpected. On this occasion the unexpected duly happened and provided a challenge which lasted well after the show.

It was a tradition at the show for FUW leaders to pay a courtesy visit to the rival Union's stand in a public display of harmony, maturity and good will. Invariably the recipients of these good wishes paid a reciprocal visit and both sides presumably went home satisfied that they had made a contribution to agricultural unity. In essence nothing could be further from the truth although in 1978 Tom Rosser was the chairman of the Welsh NFU and I believe he adopted a sincere stance towards the FUW.

My cynicism in regard to these visits was based on an unpleasant incident some years earlier when I accompanied the Union leaders to the NFU stand. As usual one was introduced to the gathered throng, shook hands and moved on. On this occasion the recipient of my handshake had, during the general hubub, not heard my name and then, realising I was from the 'enemy camp,' withdrew his hand quickly and said clearly, "That's the biggest lump of shit I've had in my hand today'. My hosts were horrified and apologised profusely.

The 1978 visit passed without incident and I had forgotten all about it when, at the end of the second day of the show, I received a confidential note, signed by Tom Rosser, which invited me back to see him as he had an issue which he wanted to raise with me. I assumed that one of my Press statements had upset him and readied myself to justify their contents. To my astonishment he informed me that the NFU in Wales was looking for a Public Relations Officer and asked whether I would be interested in the vacancy. My immediate reaction was one of outright rejection but he suggested that I think it over and that he would contact me after the show. I thought of nothing else for days but after 15 years of competing – and some would say harassing – the NFU, I could not envisage changing horses in mid-stream.

I mentioned my strong reservations to a close friend who considered that I should not reject a very good offer out of hand. When the telephone call eventually came I reluctantly agreed to attend an interview at a well-known Cardiff hotel. After the interview I was told that the job was mine if I wanted it. Richard Maslen, then the NFU's Director of Information, told me that the terms could be improved if necessary. My response was to say I would think about it – when I got home I had another call from Richard Maslen and as the NFU's Welsh Office was relocating to Cardiff I based my refusal on the fact that I would have to move to Cardiff. Richard Maslen said that arrangements could be made for me to work from the Union's Aberystwyth office. I said I would sleep on it. In fact, I had very little sleep that night and changed my mind several times -when the call finally came my reservations had the upper hand and I have few regrets about my final decision. They say that opportunities seldom knock twice – they did in my case but that's another story.

Celebrations!

Farm Minister John Silkin with Union President Myrddin Evans at the 1978 annual meeting, not long after the Minister announced the Union's official recognition. It was no secret that the Minister's relationship with the NFU was, at that time, at a very low ebb.

Recognition was a big boost for the Union and was celebrated at that day's meeting of Union leaders. Seen here are Evan Evans (NEM), Evan Lewis (General Secretary), the late Alcwyn James, R. ap Simon Jones, H.R.M. Hughes (Deputy President), the late Norman Fitter (Executive Officer) and Gwilym Thomas. The late Mrs Megan Davies, seated with FUW President Myrddin Evans, burst into tears when she heard the news.

50th anniversary medal recipients, presented for services to the Union. Back row: Gareth Vaughan (President), Bob Parry, Dai Jones, John Evans, Evan R. Thomas, Meurig Voyle, Gwilym Thomas. Front row: Llew Jones, Gwilym Jones, H.R.M. Hughes, Lord Morris, Dewi Thomas.

With FUW founder member Dewi Thomas of Carmarthen.

In animated conversation with the Duchess of Cornwall and FUW Deputy President Emyr Jones.

With Prince Charles – a courageous advocate of the family farm – at Powys Castle with Professor Michael Haines, formerly of the Agricultural Economics Department, University of Wales, Aberystwyth.

With very few exceptions I cannot complain about the FUW's treatment in the Press during my 38 years with the Union. The industry is extremely fortunate in having specialist, agricultural journalists who have – over the years – set a very high standard in their coverage of an industry, and issues, which are often enormously complex. Most of those covering farming in Wales attended the Union's 50th anniversary celebrations and are pictured here. The group, pictured here, includes myself with (left to right) Steve Dube (Western Mail), my successor Alan Morris (now at the Assembly), Barry Alston (Farmers Guardian), Andrew Forgrave (Liverpool Daily Post) and Bob Davies (Farmers Weekly).

The joke's on me – with Dei Tomos (BBC) and David Lloyd, former agricultural correspondent of the Liverpool Daily Post, at my retirement function.

A pleasant surprise – receiving the Sheperd's Crook award for farm journalism from FUW President Bob Parry in 1996.

With the late Roland Brooks, agricultural correspondent of the Western Mail for many years. We often had good natured arguments about FUW coverage but we remained firm friends; the farming industry in Wales lost an outstanding advocate when he died.

"With this kind of heat we had better agree."

Secretary of State Peter Walker attempted to reconcile the differences between the two Unions when he arranged unity talks in 1989. Farming News showed the Minister as a dragon breathing fire on the two sides – unfortunately the fire went out and the talks failed.

Quick as you can boyo — another telex for Mr Gourlay.

Another Farming News cartoon published during the 1989 unity talks. The Minister attempted to put pressure on both sides, particularly the NFU President, Simon Gourlay. Once again the NFU refused to co-operate in establishing an independent Union in a federated relationship.

Aberystwyth artists and cartoonist Hywel Harris co-operated eagerly in producing cartoons for publication in the Union's journal, Y Tir. He carried out my instructions to the letter in emphasising the dominance of the London leaders of the NFU when unity talks collapsed in 1989. NFU President Simon Gourlay is depicted pulling the strings of a dummy on his knee shown as 'Welsh NFU'.

Another cartoon shows a bewildered NFU President surrounded by Welsh farmers demonstrating outside a store selling French goods. At this time French farmers were attacking supplies of Welsh lamb in France.

Says his name is Gourlay – never heard of Cig Oen Cymru!

Two county veterans, Meurig Voyle (left) of Denbigh, noted for his warm welcome to the Royal Welsh Show stand and the late Walter Rowlands of Glamorgan.

Headquarters and county staff in 1997.

Horns of a Dilemma!

The FUW's recent campaign to draw attention to the animal health problems posed by the importation of foodstuffs by overseas travellers brought back memories of a dilemma I faced back in 1993 when I was invited to represent Welsh agricultural journalists on a tour of Scandinavian countries which were, at that time, considering joining an enlarged Common Market. It was an offer I couldn't refuse – I had never been to Norway and Sweden and the visit raised intriguing questions relating to farm support in the Less Favoured Areas – one could not think of anything less favoured than Norway's prolonged and hard winters, necessitating livestock being kept indoors for long periods! As far as Norway was concerned the issue never arose – they did not enter the Common Market.

The fact-finding tour turned out to be an exhausting flying visit to farms, meetings with farming and government officials in various parts of Denmark, Norway and Sweden. 'Flying' is a misnomer – apart from two flights most of the time was spent travelling along icy roads in a desperate attempt to keep to a demanding schedule. During these long journeys I and my fellow travellers – from Ireland, Ulster, Scotland and England – took it in turns to talk to the driver in order to keep him awake.

Most of the road journeys were in Sweden which is heavily forested and during my 'stint' of conversation with the driver he mentioned the dangers posed by elk which wandered across forestry roads and sometimes stood in the middle of the road to watch the dodging traffic! They are big animals and not to be trifled with. The conversation turned to elk meat which, I was told had a distinct flavour which I must try. I thought no more of this conversation until a few evenings before the end of the tour when the driver proudly presented me with a heavy plastic bag containing a large portion of elk meat! It was a kind offer I could not refuse – it was also a package which I could not bring back with me to the UK. The meat remained in the hotel room's refrigerator until I finally persuaded the hotel chef to take it off my hands.

Membership – Think of a Number

The respective membership figures of the FUW and the NFU in Wales have always been hotly disputed and the subject of claim and counter claim over the years. The 'numbers game' started as soon as the FUW was established with the new Union claiming that members were leaving the NFU in droves. The NFU countered by claiming that, for example, only nine members had left its Cardiganshire county branch to join the new Union in the first few months and that it would not last long.

As early as December 1955 the Western Mail published its controversial cartoon by J.O. Walker showing the 'new born FUW' out in the West Wales snow, calling for its mother – the NFU – housed in a warm and comfortable cowshed. The caption read 'He'll soon came back to mother'.

In the heat of the early exchanges both sides exaggerated their numerical strength – not long after its establishment J.B. Evans claimed that the FUW would soon have 15,000 members. FUW President Ivor Davies claimed that the Union's membership had passed the 10,000 mark in the first year. The reality was that membership was well below this figure in the Union's first two years – after the initial flush of success the Union found it difficult to attract new members in the numbers that it had hoped for. The 2000 new members enrolled in 1958 was regarded with considerable concern by FUW leaders and various recruiting targets were set for each county branch in an effort to attain its annual total target of 4,800 new members – a remarkable target in today's terms.

The new recruiting drive was launched after a highly respected member of the Union's Finance Committee, Captain H.R.H. Vaughan, said that it was perfectly clear that the actual total of members paying subscriptions was nowhere near the total given by Union officials from public platforms. Concern was also expressed at the number of members paying less than the standard rate.

In 1963 another prominent member of the FUW Finance Committee, J.E. Wright, complained that the Union seemed to have lost some of its spirit and consequently its appeal. D.J. Davies warned that an organisation which falsified its figures only postponed the evil day.

It took 20 years before the Union released accurate, audited figures of Union membership. Secretary of State John Morris had demanded an audited report on Union finances and membership to support the Union's case for official recognition. The 18 month survey, submitted in 1976, showed that the Union had 12,713 members and that the majority were, as expected, located in the western and central areas of Wales. The Union dominated counties like Anglesey, Merioneth and Cardiganshire.

FUW membership figures continued to fluctuate after that and peaked at over 14,000 in the 1980s before going into a slow and gradual decline as a result of increasing economic pressures on the industry which were particularly severe on small, family farms which represented a significant percentage of FUW membership. Despite occasional recoveries the sharp fall in farm incomes and farmer numbers of recent years have contributed to a further erosion and membership is now estimated to be considerably below the 10,000 mark.

The same scenario has taken its toll on NFU membership – the most recent administrative review has resulted in the closure of its two regional offices in North and South Wales as a result of falling membership and the subsequent need to cut costs. They have been replaced by one main office at Llanelwedd.

The NFU's consultation document 'Growing on Success – Tomorrow's NFU' emphasised the importance of attracting 'Countryside' non-farming members to compensate for the increasing erosion of its farmer members who were the victims of economic pressures.

By 2000 the Farmers Guardian estimated that 85,000 Countryside and 'professional' non-farming members had been recruited – in 2001 Director General Richard Macdonald put the total figure at 'almost 150,000'.

The Farmers Guardian estimated that in 2000 the NFU had 60,000 farmer members in England and Wales and added: "A few years ago it was nearer 85,000." In 2006 it reported that the NFU had lost more than 2000 members in 2005 and that its farmer members totalled 57,977 in England and Wales. Given this rate of erosion and the severity of the economic climate total farmer membership could be down to around 50,000 by the end of the decade.

The NFU in Wales claims 15,000 members but this is believed to include a significant number of non-farming members. The FUW recognises the importance of this alternative source of income but has not been as successful in exploiting its potential. Some farmers have pointed out the dangers of this policy in that opposing factions could infiltrate and seek to influence its policies.

United We Stand ...

The history of the FUW has been punctuated by abortive attempts to heal the breach with the NFU. The initial attempt occurred as early as the first few months of the FUW's establishment when prominent individuals in West Wales sought – without success – to pour oil on troubled waters. Two years later came the first controversial and ill-fated meeting involving leaders of the two Unions when J.B. Evans and D.J. Davies met NFU President Sir James Turner and representatives of the Welsh NFU in London. During that meeting D.J. Davies made a strong appeal to Sir James to recognise the problems, and the depth of feeling in Wales, and urged him to act 'in a bold and statesmanlike way'.

The appeal fell on deaf ears and the scene was set for over half a century of division and in-fighting in an industry which suffered strong economic pressures and experienced sweeping changes following the UK's entry to the European Union, when farming entered a new and more competitive era.

Individual members of the NFU spoke out courageously on the rift – in 1965, R. Wheelock of Monmouth campaigned for an end to the division in the industry through the medium of the Farmers Weekly. He appealed for an early termination of the bitter hostilities which bedevilled the Welsh farming scene and added that to deny Welsh farmers the right to organise their own Union was arrogant stupidity.

The following year, NFU President G.T. Williams met FUW leaders at Llandrindod but the talks foundered when it became clear that what was on offer was a return to the NFU.

The next initiative came from a meeting of the Pembroke county branch of the FUW in 1969 when local leaders of the Union met their opposite numbers in the NFU to discuss what was feared to be a looming crisis in the milk industry. However, during the meeting the agenda was extended to discuss unity. The chairman of the FUW county branch, Goronwy Griffiths, subsequently reported that the joint meeting had agreed that they should strive to achieve unity and that they should review progress at a second meeting.

However, when the news of the meetings leaked out to FUW headquarters at Aberystwyth, concerns were raised about Pembroke's action. Despite reservations, FUW headquarters sanctioned further meetings but the momentum had been lost and hopes of further progress were killed off by a statement from the local NFU pledging allegiance to London.

A few years later – in the mid 1970s – a leading member of Brecon and Radnor NFU, and chairman of the Welsh Agricultural Export Council, Austin Jenkins, made a number of attempts to settle the differences between the two organisations and advocated that a federal system of UK representation would be one solution to the problem. In 1976 the Farmers Guardian described his decision to withdraw as a mediator because of ill-health as "a major blow to the whole of Welsh farming."

The leading article in the Farmers Guardian added: "A month ago Mr. Jenkins suggested that the main factor hindering progress was the reluctance of some leaders to surrender present positions and prestige in their respective organisations. If that is so – and Mr. Austin Jenkins has been in a better position to know than most – it is time everybody involved took a long, hard look at himself. For in the end it is Welsh agriculture that matters more than individuals."

The Farmers Guardian leader added that there was no reason why a separate Welsh Farmers Union should not be set up. Both Unions would contribute to it but after an initial period of establishment, national elections could decide who was to lead and who was to serve.

Politicians of all political colours have added their voices to those calling for an end to farming's rift, including Minister of Agriculture Peter Walker, Secretaries of State Nicholas Edwards (Lord Crickhowell) and Gibson-Watt, while a number of Labour Ministers – e.g. Lord Cledwyn, who acted as a mediator at one meeting – and John Silkin joined Secretary of State John Morris in appealing for unity. After the FUW's recognition in 1978 John Morris had hoped for cooperation between the two Unions – not only were his hopes dashed, but worse still, the NFU refused to attend Ministerial meetings with FUW delegates – a development which angered John Morris who described the situation as 'nonsensical'.

A leading article in the Western Mail commented: "There can be no justification in having two Unions. Both the NFU and FUW must strive towards unity... perhaps the best solution which could accommodate the aspirations of both Unions would be the creation of a Welsh Farmers Union joined to the NFU in exactly the same way as the Scottish and Ulster farmers unions which maintain their separate identity while being joined in a federal structure with the NFU in England and Wales."

Despite evidence that members of both Unions were in favour of an end to the discord, the message was like water off a duck's back – an appeal by FUW President Myrddin Evans for both Unions to poll their members on the issue failed to get any response.

Nevertheless there were fresh hopes when, after the refusal to share Ministerial meetings with the FUW, NFU Welsh chairman Tom Rosser indicated that they were prepared to discuss the issue. Once again the NFU terms represented a return to the NFU and although a prominent group of Young Farmers Club members added their support for a settlement, both Unions issued a joint statement in May 1979 stating that the talks had foundered on the FUW's demand for a Welsh Union in a federated UK structure. The statement added:

> *"We wish to emphasise that the talks have been conducted in a most cordial and friendly atmosphere and we are now determined to work harmoniously within Wales in the best interests of the agricultural industry."*

Again there was a sting in the tail which quickly banished any hope of harmony – while the FUW delegates spent the remainder of the day socialising with their counterparts at the NFU's Cardiff headquarters, their hosts issued another statement blaming the FUW for the failure of the negotiations and accused the FUW of being 'separatist and insular.'

The stage was set for a continuation of the corrosive war of words and the two sides were not to meet for another ten years when Peter Walker's Welsh Office Agricultural Secretary, John Davies, brought the two sides together once again in Cardiff. Despite a promising start, and John Davies' best endeavours, the talks collapsed once again – for the same reasons as the previous encounter.

On this occasion the FUW delegates were mystified by the initial progress and the sudden collapse of the talks at a subsequent meeting some two months later in December 1989. Prospects for a settlement had appeared to be favourable – the NFU was reported to have financial problems and there were controversial proposals to restructure the Union, including the revision of its county structure in Wales. One North Wales NFU County Secretary admitted that a large number of his members were strongly opposed to the reorganisation plan and he feared they would go over to the FUW.

Two days before the Cardiff confrontation of the two sides, led by the NFU President Simon Gourlay and FUW President H.R.M. Hughes, a poll of Anglesey farmers on S4C's 'Ffermio' programme showed that 66 per cent supported the FUW's policy of autonomy and federation. Over 50 per cent of the farmers polled – including 15 per cent of NFU members – considered that the FUW best represented their interests.

During previous talks the FUW delegates had the distinct impression that they had failed because of the influence of die-hard opponents in the Welsh NFU. FUW President Myrddin Evans blamed this influence on the failure of his talks with NFU President Henry Plumb who supported the principle of federalism. But in 1989 the FUW delegates considered that the roles were reversed and that the Welsh representatives, who included the Welsh Council chairman, were subservient to their London leaders. The NFU team was led by President Simon Gourlay who was considered to be a 'hawk' while it was felt that Deputy President David Naish had clearly indicated his – and London HQ's views* – when, some months earlier at a meeting of the NFU Welsh Council, he had asked

* The London headquarters of the NFU moved to Stoneleigh in 2006.

whether it was really sensible for a separatist organisation like the FUW to represent Welsh farmers when most of their suppliers lived and worked in England! He also accused the FUW of setting preconditions to unity talks.

Nevertheless, when mediator John Davies announced that the talks had failed after what was considered to be early progress, it came as a bitter disappointment to FUW President H.R.M. Hughes and a surprise to expectant observers. In a subsequent statement H.R.M. Hughes said:

> *"Despite the appeal made by the Secretary of State and the efforts of an independent chairman, we have failed to find a formula which would enable us to end the division – and some would argue, the duplication of effort – which has characterised the agricultural scene in Wales for the last 33 years. I want to emphasise that the FUW went into the talks with hope and determination to resolve the differences between us."*

Once again the NFU post-mortem on the failure of the talks allocated the blame on the FUW and condemned my action in issuing a Press Statement soon after John Davies' announcement. My statement regretted the failure and included the comments made by the FUW President featured above. The NFU suggested that the immediate production of the statement indicated that the FUW had anticipated the break-down of the negotiations! There were some red faces when I was able to prove that I had, in fact, prepared two Press Statements, one regretting the failure of the talks and another – which never saw the light of day – welcoming and celebrating a successful outcome. The production of two statements was the only way of overcoming my isolation from administrative facilities in London. In fact, my hopes of a settlement had been raised by the early progress in the negotiations and it was with a heavy heart that I issued the 'unsuccessful' version and disposed of the alternative version.

My acute sense of disappointment influenced my decision to highlight the perceived failure of the Welsh delegates to recognise the advantages of one Union at a time when agricultural devolution was around the corner. I asked a well-known artist, Hywel Harris, to produce a cartoon showing the Welsh NFU chairman as a puppet on the knee of 'ventriloquist' Simon Gourlay, proclaiming in a caption 'We are the independent voice of Welsh farmers.' I used the cartoon on the front page of the FUW journal, much to the fury and indignation of the NFU's Parliamentary Committee former chairman, John Hooson of Denbigh. He demanded that I print his letter of protest because, he said, the cartoon implied, quite unambiguously, that Welsh NFU members were puppets manipulated by London. I was more than happy to accede to his request in order to reiterate the whole point of the cartoon!

The balance of representation on the NFU side during the negotiations was heavily weighted against any compromise and, in hindsight, one understands H.R.M. Hughes' conclusion that he did not believe that the talks stood any chance of success. The issue was not raised for some time after this failure – in the early 1990s a leading member of the NFU, Hywel Richards of Criccieth, did suggest that independent arbitrators should be appointed to work towards a settlement. Neither side reacted with any enthusiasm to the suggestion.

The establishment of the Welsh Assembly has raised fresh hopes that unity is not a lost cause. The Assembly, which has considerable influence over agricultural policies in Wales, has brought the two sides closer together – particularly during periods of crisis like the 2001 foot-and-mouth outbreak and, more constructively, the reform of the Common Agricultural Policy (CAP). The Assembly has established a new avenue of representational cooperation. NFU and FUW leaders were members of the 2001 Farming Futures Group which produced the blueprint policy document 'Farming for the Future' and were co-signatories to a document which stressed that the future of farming in Wales would be guided by a distinctly Welsh strategy.

Welsh Rural Affairs Minister, Carwyn Jones, emphasised that collaboration and cooperation were the way forward for farming in the future. He commented: "The industry must change to challenge the gradual decline over the past decade – working together we can sustain agriculture as important to Wales."

Nothing illustrated the cooperation of the two Unions more effectively than the Luxembourg summit of 2003 when the NFU Welsh chairman, Peredur Hughes, and the FUW President Gareth Vaughan, joined together with Welsh Minister Carwyn Jones and UK Agriculture Minister Margaret Beckett to put the case for Welsh farming in the run-up to the all-important revision of the EU's Common Agricultural Policy.

Wales subsequently, unlike England, was successful in achieving the introduction of a new Single Farm Payment (SFP) based on the historic system – a historic victory in itself. FUW President Gareth Vaughan commented: "We are already seeing the benefit of devolution in agriculture... farmers over the border in England view our settlement with envious eyes."

Joint Union representations have been made on a number of other important issues and both Unions have called for the Welsh Assembly to be granted more powers. Referring to the additional powers conferred by the Government of Wales Act, NFU Cymru chairman Dai Davies stated: "These powers mean that NFU Cymru, in conjunction with government, can now build on what has already been achieved and continue to develop Wales' own distinctive agricultural and rural policy."

The 2001 foot-and-mouth outbreak proved to be a catalyst for cooperation and joint action at local and national level. Both Unions came together in North Wales and cooperated effectively to produce a rescue package for the important Bryncir market which was threatened with closure. FUW chairman Glyn Roberts commented at the time, "The FUW and NFU worked together to help devise the rescue package."

In recent years new voices have questioned the sanity of the current position in which two organisations feed off an ailing industry. The most outspoken – and courageous – appeal for unity came from NFU Wales chairman Hugh Richards who told the Welsh council of the Union in a valedictory address that agricultural organisations should be working together to find a way forward.

He added: "As far as the NFU is concerned we should be looking at a federal structure and a means of getting everybody in the industry working together in a common direction... we have to take the lead and start bringing people together... from Scotland, Northern Ireland, England and Wales, as well as organisations ranging from the Country Landowners Association, National Sheep Association, National Beef Association and even the Farmers Union of Wales. Let us try and find a way forward together because we are getting fewer in number as an industry. A federal structure, an association of agriculture must be the way forward."

These, and subsequent views expressed by NFU Cymru leaders, suggest that there there has been a seismic shift in attitudes since the NFU refused to share the negotiating table with the FUW over a quarter of a century earlier. Former NFU Cymru President, Peredur Hughes, acknowledged this in a television interview and has added that NFU Cymru is 'completely independent in terms of deciding agricultural policies in Wales'. He added: "I would like nothing more than for Welsh farmers to be able to speak with one voice but at the end of the day that is a decision which can only be taken by the members of each Union" – a possible reference to the acceptance of a poll of members.

A more academic but equally frank appraisal of the agricultural scene came from Professor Gareth Edwards-Jones, head of Bangor University's Agriculture and Land Use department who told delegates at the 2005 NFU Cymru annual conference, that the industry was hampered by having two unions. Unity was the only solution, he said and added: "It has to come from you and them ... it will upset people but it is the only answer."

Despite its prominence the NFU has had its critics in recent years and there have been advocates for more militant action, particularly in the milk sector. Economic pressures have reduced farmer numbers and weakened the industry's political influence, particularly at Westminster. "UK farmers have shrunk in numbers and shrunk even faster in political influence ... the farming vote ceased to be of any political influence many years ago" (Farm Law).

A letter in the Farmers Guardian from a NFU Council delegate appealing for unity added "Like it or not, the NFU no longer stalks the corridors of power – instead it patiently waits in the lobby of influence with a myriad of other pressure groups." The same edition of the Farmers Guardian emphasised in its editorial "If farming has one weakness which holds it back time and time again it is the inability to stand together."

The FUW is, on the other hand, superficially successful – it has achieved the key goals it set itself in 1955 – it can pursue policies of importance to Welsh farmers and since recognition it has access to the industry's policy makers. However, the reality is that it presides over an industry in serious decline. Membership numbers have fallen substantially, reflecting the economic pressures and the flight from farming. While acknowledging the Assembly-led success in the Single Farm Payment arrangements, the Westminster government still represents UK farming in Brussels and presides over major farm policies – both Labour and Conservatives are, for example, committed to CAP changes.

The FUW can be proud of its achievements – it has survived the sacrifices and the crises of its early years and it would be a tragedy if it could not now meet the aspirations of its founder members and early leaders by failing to play an important part in the difficult period ahead when changes in support measures and new members of the European Union start to make their presence felt.

Over the years every leader of the FUW has initiated attempts to seek an honourable settlement to the damaging division in the Principality's farming industry. The importance of unity has been recognised by recent and current leaders of both Unions. As far as the FUW is concerned, FUW President Gareth Vaughan stated at the Union's 50th anniversary annual meeting that the Union was committed to cooperation with other organisations and would continue to form new alliances.

Following that statement the FUW joined with the NFU and other organisations in the promotion of the livestock export trade and, subsequently a badger (TB) survey and, following a lead set by the FUW Milk Committee chairman, Eifion Huws, a campaign drawing attention to the on-going problems in the milk sector.

Against this background it came as a surprise when NFU Cymru President Peredur Hughes stated at the end of his period in office that his biggest disappointment had been a failure to bring about a merger of the two Unions and that 'meaningful' discussions between the two organisations had been halted by the FUW. There are sources which suggest that the abortive meeting referred to by the NFU Cymru leader took place in London but no details are known and the FUW has neither confirmed the event nor indicated if any progress was made at the negotiations. There is also speculation that a similar approach from current NFU Cymru leader Dai Davies has suffered the same fate. If this is the case, rank and file FUW members are unaware of it or the reasons why the negotiations failed.

It has been suggested that the FUW's attitude reflected the fact that the NFU's overtures occurred at around the time that the Union was celebrating its 50th anniversary and that unity meetings could have undermined the various celebratory events which marked the occasion.

Despite conciliatory statements and a readiness to cooperate on issues of national interest, the FUW's current outlook on unity appears to be 'so far, and no further'; indicating a marked reluctance to embrace the concept of unity. The 'cloak and dagger' treatment of this issue not only represents a reversal of previous policies but – in failing to notify its grass-roots members of these developments – has undermined the Union's once much admired democratic process.

A new and dramatic impetus to the cause of unity occurred at the 2007 annual meeting of NFU Cymru when – led by Hugh Richards – Carmarthen and Glamorgan branches of the Union urged the NFU to move to a federal structure. He told the meeting that, increasingly, decisions on Welsh agriculture were being made in Cardiff, removing the relevance of the 'NFU's English arm'.

A large majority vote instructed the Union's Welsh management board to discuss the proposal and report back "at the earliest opportunity". It was emphasised that the initiative would not prejudice relationships with central NFU and would not involve a merger with any other organisation – a thinly veiled reference to the FUW.

On the evidence of this break-through and recent conciliatory approaches, it would appear that the most important of the FUW's basic aims – and a major hurdle in previous inter-Union negotiations – has been conceded. Its efforts to establish a system of federal representation have been accepted – removing the last obstacle to unity. Astonishingly there now appears to be some doubt as to whether even this concession will bring the two sides together. It has been suggested that a mediator is involved – if so it seems to have done little to speed the healing process!

The FUW's founder members and the Union's previous leaders would be perplexed and concerned by this sluggish response to the NFU initiative which appears to achieve the objectives which they listed as their fundamental aims when the Union was established and have been fought for, and pursued, ever since.

D.J. Davies, the first chairman of the FUW's Policy Committee stated categorically at Dihewyd in 1957: "Once Wales is conceded the right to speak for its own farmers, then, when that day comes, I would say – heal the division at once! Let both organisations merge ... the FUW was formed to get for Wales, the same, nothing more, as other countries of the UK".

Numerous attempts to end the division have foundered on the issue of independence and a federal structure. FUW leaders Myrddin Evans and Glyngwyn Roberts rejected NFU unity overtures at a Llandrindod meeting as early as 1966 when it became clear that the NFU would not give Wales the same status as Scotland and Ulster.

In a campaign for new members the FUW's Membership Guide (circa 1970) stated categorically: "The FUW exists to serve farmers and growers in Wales. We think they are entitled to an organisation which will look after their particular interests. Scotland and Ulster have their own Unions, independent but linked with the UK organisation. In this way they can make a valuable contribution to UK policies and yet pay special attention to the problems which exist in their own countries. THIS IS EXACTLY WHAT THE FUW SEEKS TO ACHIEVE."

FUW President Myrddin Evans reiterated the principle in a major speech in 1972 when fresh efforts were made to bring the two sides together. Austin Jenkins, a prominent member of Brecon & Radnor NFU, who attempted to act as a mediator in the 1970s stated in a letter to Myrddin Evans (July 1976): "There will have to be give and take on both sides eventually but that will only be after the principle of federalism has been agreed."

The federal approach was prioritised in all the representations on the FUW's part during unity negotiations over the years. The statement issued to the Press after the unsuccessful 1979 unity talks emphasised that they had foundered on the FUW's specific demand for the establishment of a Welsh Union in a federated UK structure.

Over the years FUW leaders and individual members have made enormous sacrifices for an all-important point of principle and for what they perceived was an injustice to farmers in Wales. It is inconceivable that this principle is now to be cast aside and that, when Welsh agriculture faces its toughest test, efforts to heal the rift should be ignored.

Against this pessimistic background it has even been suggested that a 'trial marriage', which puts the interests of the industry first and foremost, may still be possible. Given the will to succeed, allied to courageous leadership and statesmanship, it may still be possible for both Unions to retain domestic control and erect a framework of active and regular cooperation over, say, a two year experimental period which would gradually reduce and, possibly, eliminate expensive duplication. This would entail regular consultation on policy issues.

Some years ago, Henry Fell, the Yorkshire sheep farmer and respected observer of the industry's fortunes – alarmed by the industry's decline and representational weakness – urged the various producer groups in the industry to join together and emulate the Confederation of British Industry (CBI) by establishing a Confederation of British Agriculture (CBA) to fight for the farming industry.* As the years have passed and the economic pressures on the industry have increased, others have

---

* The CBI was established in 1965 as an independent non political voice to articulate the needs and problems of business to government. It is organised into regional Councils and a national Council, chaired by a President.

echoed his proposal which would establish a loose but effective association of farming organisations dedicated to the industry's economic interests.

Henry Fell commented at the time: "Our problem is one of weakness, a transparent weakness that is nothing less than a gift to our main customers, the supermarkets. I am aware that the CBA proposal may represent an over-simplification and needs detailed examination. I also know it will hurt individual pride – so be it. The desperate need is self evident – do nothing and the consequences don't bear thinking about."

This is the challenge facing leaders and members of both Unions. Previous differences must be cast aside and the emphasis placed on the best interests of members at a time when farming faces a difficult and very uncertain future. The founder members of the FUW overcame all odds to build an organisation from nothing. If the industry is to survive and flourish current leaders and members must show the same courage and statesmanship.

Farming – Stand and Deliver?

Welsh agriculture is at a cross-roads. How many times have we heard that? I worked with and on behalf of farmers for 38 years and became familiar with the taunt that farmers are never satisfied with their lot. In fact, this is not the case – there have been times when farming has flourished and farmers have acknowledged the fact. However, despite human endeavours and skilled management, the best laid plans can go wrong and as farming's fortunes can depend on factors outside their control – the weather, animal health and political regulation – farming can be a frustrating and debilitating experience.

In Wales, farming is also governed by limitations imposed by the terrain and quality of the land.

In recent times, costs of production have spiralled and fears about the very future of the industry have reached a crescendo, particularly in relation to the milk sector. The opinions, views and comments which are noted here illustrate the fears of a varied cross-section of respected commentators and leading figures in the industry.

In July 2006 the Royal Agricultural Society published a devastating report: 'Differentiation – A Sustainable Future for UK Agriculture' which highlighted farming's problems and raised serious doubts about its future. Presenting the report, the President of the Society, Sir Stuart Hampson, chairman of John Lewis Partnership – which includes the Waitrose chain of supermarkets – who commissioned the report, warned that a "deep-seated crisis" afflicted farming, leaving the industry perilously close to collapse.

The report concluded that the problems represented "not just another crisis" but a turning point that would have profound implications on the countryside, primary food manufacturers and consumers, unless urgent remedial action was taken.

The findings of the report included the opinion that a viable farming sector was "very close" to being lost forever. The report stated that the UK was only 60 per cent self sufficient in food and that it imported 27 per cent of the food which could be produced at home – a figure that had doubled during the previous decade.

In Wales, economic pressures have transformed the agricultural scene; the milk sector, once the prosperous cornerstone of the industry has declined rapidly due to increased costs of production and a retail sector dominated by the supermarket chains. Membership of the European Union resulted in the loss of the Milk Marketing Board and the introduction of quotas to control over-production. The once often quoted original aim of the Common Agricultural Policy (CAP) – 'to ensure a proper standard of living for farmers by increasing earnings' has had an increasingly hollow ring for milk producers in recent years.

In 1994 there were 5,363 holdings in Wales with dairy cattle – by 2006 the number of dairy farmers had fallen to 2,500 and the sector has been declining at an annual rate of seven per cent. Some fear that the exodus could accelerate in the near future, leaving the sector in the hands of a small number of large production units. UK milk production is running at less than 70 per cent self-sufficiency and steadily declining, raising fears of liquid milk shortages in the near future.

There were hopes that the Competition Commission's inquiry into the supermarkets and grocery sector might provide some measure of relief. NFU Deputy President Meurig Raymond, who with his brother Mansel, NFU Cymru's dairy board chairman and a First Milk director have one of the largest dairy enterprises in Wales, commented that while he detected some realisation of the sector's problems, producers were at a cross-roads. He added: "Many have already decided to get out, output volumes are falling and will probably fall even further."

The scale of the farming recession in Wales was highlighted by 2006 Welsh Assembly estimates of farm incomes which indicated a huge drop of £44 million – 29 per cent; from £152 million in 2005 to £107 million in 2006. As generally feared the milk sector suffered the most with the value of milk and milk products falling by £23.9 million. Sheep and lamb, and poultry output fell by £8 million and £7 million respectively. Input costs during the year rose by £24 million to £771 million due mainly to higher fuel prices. Cattle and calf output increased by £18 million.

In totality the 2006 income figure was 11 per cent lower than that of 2001.

The figures contradicted DEFRA figures of three months earlier showing a seven per cent rise in UK incomes.

The consequences of the current recession in the farming industry were accurately forecast as far back as 1999 in 'Rural Wales'. the economic review by the Rural Partnership:

> *"Farming has a strong tradition in Wales and family farms have been at the heart of that tradition providing direct income and employment, supporting rural economies, managing the countryside, contributing to community life and helping to sustain the Welsh language and culture.*
>
> *The family farm is at the heart of many of rural Wales' communities. However, such farms are under severe threat. Farm employment is declining rapidly, many farms are being lost by amalgamation, young members of farming families are opting out of the industry in the face of an insecure future and the average age of Welsh farmers is rising. The current agricultural crisis has made a bad situation dramatically worse, driving many farmers into serious debt or to sell holdings which may have been in the same family for generations. There are compelling economic, environmental and social reasons for supporting the family farm. New economic opportunities are emerging based on quality products which add value and promote collaborative marketing. The current farm structure could also be vital to the conservation of the Welsh landscape."*

A correspondent to a national newspaper put it succinctly: "The disastrous foot and mouth outbreak of 2001 highlighted the way the farming industry provides the base for many other industries and how they were adversely affected because of farming's closure. A country that ignores the plight of the farming industry does so at its peril. British farmers produce the best food available. It is inexpensive to buy and it has quality and traceability second to none. There should not be this disparity between the price paid to the primary producer and the price the retailer charges the buying public."

In 2001 the Welsh Assembly's report on the key strategy policy for farming in Wales 'Farming for the Future' emphasised that if Welsh farming continued to try to compete on price alone, as a producer of agricultural raw materials, the future of the Welsh family farm and Welsh rural life were grim. The report spelt out the bleak and uncompromising reality of the future: 'Rural prosperity will depend vitally on generating new employment and business opportunities OUTSIDE agriculture'. The report added that the competition that Welsh farming was already facing in commodity markets would increase still further when the countries of eastern Europe 'gear up their agriculture – they have enormous agricultural potential, cheap land and cheap labour'.

Central government has shown little sympathy for the industry's plight and many farmers blame Ministers and the Department for Environment, Farming and Rural Affairs (DEFRA) for a haphazard approach to payment deadlines and pressing policy issues. Commenting on DEFRA's annual survey of the industry 'Agriculture in the UK 2005' the Farmers Weekly's respected columnist David Richardson observed gloomily "The record of agriculture as portrayed in this, DEFRA's most important annual survey, is one of almost unremitting decline."

Farming's luke-warm attitude towards Tony Blair's Labour government was not surprising in view of the indication in its 1990 election manifesto that DEFRA's priority in the future would be consumers and not producers. Farmers in Wales escaped DEFRA's handling of the Single Payment Scheme and it is only more recently that the harmony that existed between farming organisations and the Welsh Assembly has suffered from the proposal to cut Tir Mynydd (hill farming) payments.

Central government continues to pour fuel on the flames of disillusion by advocating further changes in Common Agricultural Policy support and other restrictive measures which would result in increased costs. One DEFRA minister even told the Labour Party's 2006 annual conference that the dairy industry needed a 'massive shake-out' to rid itself of inefficient producers.

An enforced and accelerated shake-out could have serious consequences for farming, the economy of our rural areas and the environmental and the cultural qualities which give our rural areas a unique identity. Rural areas play a significant role in the tourist industry which produces annual revenues of around £2.5 billion – £6 million a day – half of which is generated in rural Wales.

Many rural areas have become more isolated from the services taken for granted in our towns and cities. Proposed changes in medical and social services, the closure of rural post offices and the absence of social centres have had a negative impact on rural life. Young couples are unable to compete with retired incomers for housing and are forced to leave for greener pastures.

Young farmers have reacted as one would expect to the pressures on the industry. A number of surveys have indicated that an increasing number do not regard farming as a full-time career. With the average age of farmers increasing generally – in Wales in 2006 it was 50+ – there is a danger that farming will not attract the brightest entrants and that our rural areas could become dormitories for the elderly.

Twenty-five years ago the Young Farmers Club movement of England and Wales had over a thousand clubs and more than 50,000 members. Today the membership is down to around 21,000 and the organisation is running at a financial loss. Welsh clubs have been more successful at retaining members. Student applications at agricultural colleges and interest in land-based studies have declined, resulting in a rationalisation of courses. While student applications to universities generally fell by 3.4 per cent in 2006, those for land-based courses fell by 8.8 per cent.

FUW President Gareth Vaughan voiced the concerns of many when he commented: "My greatest concern is for our young people. We need to retain our young farmers in order to contribute to the economic, cultural and social fabric of our rural areas. I fear that more and more young people are choosing to follow careers away from the farm."

All this bodes ill for the future of farming, the countryside generally and what is referred to euphemistically as the 'Welsh way of life', a term which applies in particular to Welsh speaking areas and the culture based on the language. The chairman of the Royal Welsh Agricultural Society, Alun Evans, told members at the 2006 Royal Welsh Show that the countryside faced significant changes – cheap foreigh food was being flown across the world and was replacing home-grown supplies.

He added: "Only a prosperous agriculture can care for the countryside and retain those qualities which we value so highly."

A more recent NFU report 'What Agriculture and Horticulture Mean to Britain' states that in 1996 UK farmers and growers accounted for 60 per cent of the country's total food supplies but that since then overall self-sufficiency had dropped by 18 per cent.

Sir Winston Churchill once commented: "Fifty million people, dwelling on a small island, growing enough food for only, shall we say, thirty million, is a spectacle of insecurity which history has not often seen before."

Prince Charles has also warned of the consequences of not being self-sufficient in a world faced with uncertainty. He told a NFU conference: "Food security takes on an even greater importance in an ever more uncertain world – we must ensure that the country is able to supply its own needs as much as possible."

The difficulties facing hill farmers were highlighted in an earlier report by Professors Peter Midmore and Richard Moore-Colyer of UWA Aberystwyth (Cherished Uplands – Brecknockshire Agricultural Society – Institute of Welsh Affairs) which emphasised that most hill farmers were dependent on income from outside farming and that despite their commitment to farming, the uplands were at a 'tipping point'. The report warned that without favourable support a 'worst case' scenario of land abandonment and widespread environmental degradation was possible.

British Veterinary Association President, Freda Scott-Park, in her valedictory address to the Association, referred to the Association's open letter to Prime Minister Tony Blair in which she warned of a 'melt-down' on the UK's family run dairy farms and emphasised that unless something was done to improve farmgate prices, they would witness farming's demise.

Farming's furtunes are forecast to improve in some sectors in the foreseeable future but it remains to be seen how many farmers can survive until then. A minority have demonstrated that determined entrepreneurial and management skills can overcome serious odds. A wide variety of diversified enterprises have brought their reward.

Hopes have been raised by the success in developing new markets abroad. Red meat exports to Italy, for example, rose by 60 per cent in 2006 while there were also significant increases in lamb exports to Continental countries. Hopes have also been boosted by evidence that the Middle East could become an important, developing market.

Family farms in Wales have, in the past, been able to pull in their belts to weather a variety of pressures. However current problems are more deep-rooted and represent a more implacable scenario. Farmers have been increasingly demoralised by a burgeoning raft of rules and regulations.

It is encouraging that consumers are beginning to appreciate the dangers of importing food from countries where standards of production fall far below those governing home produced food. One national newspaper, reporting on the 2007 bird flu outbreak commented: "The truth is that very few of us know exactly what we are eating and where it came from ... From printing misleading labels to diluting or modifying the food itself, it has become more commonplace than ever for suppliers to dupe consumers."

Certainly there are solutions to farming's current decline but when one reviews the farming scene over the last decade it is difficult to avoid the conclusion that the industry's greatest need in the future is a sympathetic and supportive government.

Agriculture in Wales is at a cross-roads – the coming years will see rapid changes when only the fittest in business and husbandry terms will survive. These changes will certainly leave their mark on farming and the Welsh countryside. Whether that mark will be regarded as a welcome turning point or an ugly stain remains to be seen.